The Origins of Ancient Greek Science

This book examines the origins of ancient Greek science using the vehicles of blood, blood vessels, and the heart. Careful attention to biomedical writers in the ancient world, as well as to the philosophical and literary work of writers prior to the Hippocratic authors, produce an interesting story of how science progressed and the critical context in which important methodological questions were addressed. The end result is an account that arises from debates that are engaged in and "solved" by different writers. These stopping points form the foundation for Harvey and for modern philosophy of biology. Author Michael Boylan sets out the history of science as well as a critical evaluation based upon principles in the contemporary canon of the philosophy of science—particularly those dealing with the philosophy of biology.

Michael Boylan (Ph.D. University of Chicago) is Professor and Chair of Philosophy at Marymount University. He has delivered invited lectures in fifteen countries on five continents. Most recently, he has spoken on Greek science at the American Philological Association; an international United Nations Educational, Scientific and Cultural Organization (UNESCO) conference in Pergamum, Turkey; and at the Sorbonne (Paris IV).

Routledge Monographs in Classical Studies

1 **The Roman Garden**
Katharine T. von Stackelberg

2 **The Eunuch in Byzantine History and Society**
Shaun Tougher

3 **Actors and Audience in the Roman Courtroom**
Leanne Bablitz

4 **Life and Letters in the Ancient Greek World**
John Muir

5 **Utopia Antiqua**
Rhiannon Evans

6 **Greek Magic**
John Petropoulos

7 **Between Rome and Persia**
Peter Edwell

8 **Passions and Moral Progress in Greco-Roman Thought**
John T. Fitzgerald

9 **Dacia**
Ioana A. Oltean

10 **Rome in the Pyrenees**
Simon Esmonde-Cleary

11 **Virgil's Homeric Lens**
Edan Dekel

12 **Plato's Dialectic on Woman**
Equal, Therefore Inferior
Elena Blair

13 **Roman Literature, Gender, and Reception**
Domina Illustris
Edited by Donald Lateiner, Barbara K. Gold and Judith Perkins

14 **Roman Theories of Translation**
Surpassing the Source
Siobhán McElduff

15 **Displaying the Ideals of Antiquity**
The Petrified Gaze
Johannes Siapkas and Lena Sjögren

16 **Menander in Contexts**
Edited by Alan H. Sommerstein

17 **Consumerism in the Ancient World**
Imports and Identity Construction
Justin St. P. Walsh

18 **Apuleius and Africa**
Edited by Benjamin Todd Lee, Ellen Finkelpearl and Luca Graverini

19 **Lucian and His Roman Voices**
Cultural Exchanges and Conflicts in the Late Roman Empire
Eleni Bozia

20 **Theology and Existentialism in Aeschylus**
Written in the Cosmos
Richard Rader

21 **Rome and Provincial Resistance**
Gil Gambash

22 **The Origins of Ancient Greek Science**
Blood—A Philosophical Study
Michael Boylan

Other Books in This Series:

Childhood in Ancient Athens
Iconography and Social History
Lesley A. Beaumont

Late Classical and Early Hellenistic Corinth
338–196 BC
Michael D. Dixon

The Origins of Ancient Greek Science

Blood—A Philosophical Study

Michael Boylan

Routledge
Taylor & Francis Group
LONDON AND NEW YORK

First published 2015 by Routledge

2 Park Square, Milton Park, Abingdon, Oxfordshire OX14 4RN
52 Vanderbilt Avenue, New York, NY 10017

Routledge is an imprint of the Taylor & Francis Group, an informa business

First issued in paperback 2019

Library of Congress Cataloging-in-Publication Data

Library of Congress Control Number: 2015937666

ISBN: 978-0-415-84393-5 (hbk)
ISBN: 978-0-367-86843-7 (pbk)

Typeset in Sabon
by Apex CoVantage, LLC

Contents

Preface ix
Acknowledgments xiii

1 Blood, Magic, and Science in Early Greek Thought 1

2 The Hippocratic Period—Philosophy versus Clinical Biomedicine 25

3 Aristotle—The Origin and Function of Blood and Body Parts 49

4 Hellenistic, Alexandrian, and Roman Methodological Battles 77

5 Galen—The Grand Synthesis 109

Glossary of Philosophical Terms 137
Bibliography 141
Names and Subjects Index 157
Texts Cited Index 167

Preface

As the title of this books suggests, it is essentially a philosophical work—particularly the philosophy of science. One of the methodologies in the philosophy of science is to view the explanandum from an historical perspective using current discussions on what constitutes good scientific practice. This is termed the history and philosophy of science. In the "history" part we are situated in classics of the Western tradition. I do offer some original translations that follow my theoretical bent, but for the most part I fall in line with existing scholarship (with some noted exceptions). What is particularly original about this treatment is my use of various critical devices in contemporary logic and philosophy of science to interpret the significance of ancient biomedical texts—especially along theoretical lines. Some people claim that this is anachronistic because they do not want to engage in cross-historical analysis. Most advocates of this position are those classicists who want to keep examinations of the ancient world strictly within its own context only. However, it is my contention that the philosophy of science is a timeless realm such that what is good scientific methodology in the ancient world is also good methodology today—and vice versa. This gives a particular contemporary relevance to the study of these ancient texts. It is not *only* that we seek to know how they investigated issues concerning blood, but *also* because it lends insight to eternal truths about scientific methodology that can guide us today.[1] Thus, readers should not be surprised at my use of some modern terms such as *Homo sapiens*, biomedicine, clinical medicine, gonads, zygotes, etc. These are neither anachronistic nor positivist, but rather, an historical and philosophical device that is meant to bring ancient authors into a timeless discussion.

What I have done is to examine biomedical writers on the topic of blood in order to ferret out eternal principles that can guide biomedical science. This is an original undertaking that follows the work of Vivian Nutton, Geoffrey Lloyd, and Philip van der Eijk (from the critical, classical historiography side); Jacques Jouanna and Paul Demont (from the acute approach to language that leads to the holistic interpretations side); and Jim Hankinson (from the philosophy side).

I see several audiences for this book. The primary audience is a general reader interested in ancient medical history and the philosophy of science.

This reader should have some interest or familiarity with either (a) ancient Greek and Roman medicine or (b) contemporary debates in the philosophy of biomedical science. The secondary audience concerns specialists in either of the two mentioned areas of discourse who might wish to expand their understanding of this interdisciplinary undertaking from a viewpoint that is less familiar. The tertiary audience is advanced undergraduates and graduate students who are in programs that overlap with either of the previous categories. There is an attempt at rigor in each area with a presentational tone that is welcoming to all these audiences.

Because this is an interdisciplinary enterprise, some care has been taken to make it as accessible as possible. All Greek words are translated and then transliterated in the body of the text. Quotations in Greek only occur in the endnotes. Likewise, philosophical and logical concepts are given an accessible presentation for those readers unfamiliar with them. In order to especially help those less familiar with philosophy, I have included a glossary of some key philosophical terminology that I use in the book. It would be helpful to bookmark the glossary pages so that the reader can keep track of the terminological distinctions that I draw.

The subject matter, biomedicine, is rather broader than many standard treatments on blood that tend to focus exclusively on the history of medicine. By bringing biology into the spotlight, more general principles of science can be identified and examined.

The end product is an ambitious project that intends to appeal to a broad cross-section of readers as it seeks to expand the genre of history and philosophy through a narrow initial focus, blood, and a very broad background—how a select group of biomedical writers made theoretical choices that resonate not only in their own historical context, but still continue to be relevant to biomedical writers of today.

Key features about this book are as follows:

- A broad historical presentation on how blood was understood in the ancient Greek world for almost 900 years culminating in Galen
- Detailed presentations on key terms and concepts that have philosophical relevance concerning the understanding of how science and medicine are presented
- Presentation of many of the current discussions on these issues within the community of classical scholars and philosophers
- Adaptation of current distinctions in philosophy of science to better illumine the significance of the theories and practice of these ancient authors
- An original critical methodology that seeks ways to evaluate the effectiveness of these ancient authors in their presentation of science via the context of normative criteria developed by contemporary philosophers

It is hoped that this unique presentation will reach a wide audience, and that the interdisciplinary approach will be a broad net rather than a distraction to specialists who often view these texts more narrowly.

NOTE

1. I have taken this task to heart in my co-authored book, *Genetic Engineering: Science and Ethics on the New Frontier* (Upper Saddle River, NJ: Prentice Hall, 2002).

Acknowledgments

As in any project of this kind, there are many to thank. I would first like to thank Vivian Nutton, who has given me lengthy comments on every chapter. Next, I thank Paul Demont, Philip van der Eijk, Paul Keyser, Sylvia Berryman, Geoffrey Lloyd, and Ralph M. Rosen. I would also like to thank those participants at the UNESCO conference on Galen in Pergamum, Turkey, for their comments on versions of the early chapters and the audience at the American Philological Association session on ancient medicine who made helpful comments. Finally, I would like to express my appreciation for those who came to my graduate seminar at the Sorbonne (Paris IV) and their extended comments over several months.

I would also like to thank Laura Stearns, my publisher, who has stood by this project from the beginning, and all the production team at Routledge. Then there is my research team at Marymount: Tanya Lanuzo and Lynn McLaughlin, who assisted me in getting the best books and articles for my discussions. As always, I thank my wife, Rebecca, and my children, Arianne, Seán, and Éamon, for their intellectual acumen and loving support. None of this would happen without them.

1 Blood, Magic, and Science in Early Greek Thought

What is there about blood that makes it an essential ingredient in accounts of the human experience from medicine to poetry? This chapter will explore some of the understandings in early Greek thought about blood, the heart, and blood vessels from the time of Homer to Hippocrates.[1] Since this marks what is generally understood to be the period from pre-natural philosophy to its initial development from Thales onward, most of the comments in this chapter will be suggestive and interpreted against the background of the subsequent history of science and theoretical issues in the philosophy of science.

In order to structure this journey, it will be the strategy of this treatise to concentrate on several key passages that are illustrative of a developing biomedical understanding of the blood and its relationship to the body. For the most part, the chapter will proceed chronologically examining principally texts from Homer[2] up to the Hippocratic Writers (viewing these as our historical points of departure and arrival).

Let us begin with Homer and pre-natural philosophy. In the Homeric texts, blood is often associated with *ichor*. These are related concepts. Blood is a very basic fluid for human life. It doesn't take long for peoples prone to warfare to discover the coincidence of slashing blows that release great quantities of blood and the victim's consequent death.[3] Thus, from the outset, blood is seen to be a vital fluid. Vitality is, by its very nature, mysterious—it constitutes the difference between life and death. For this reason, ancient writers would interpret this mystery by ascribing the possession of various virtues and vices to blood along with powers of procreation and sometimes even the source of thought itself. In this survey of texts, the primary character of blood will be sketched out along with a critical analysis of the terms of these attributions.[4]

Ichor is a substance that seems to exist within blood. In some contexts the mode of this existence is as the watery part of the blood.[5] In these contexts, *ichor* is cited as providing some key material property of blood. Thus, in our exploration of blood we must also include *ichor*.

Blood is associated with courage (a virtue).[6] *Ichor* in one context is connected to the gods as being their substitute for blood (i.e., humans have blood and the gods possess *ichor*).[7] In the *Iliad* we find:

> . . . and blood immortal flowed from the goddess,
> *ichor*, that which runs in the veins of the blessed divinities;
> since they eat no food, nor do they drink of the shining
> wine, and therefore they have no blood and are called immortal.
> (V: 339–442, cf. 416)[8]

The gods are above mortals. One sign of this is their possessing only *ichor* in their blood vessels instead of a mixture of blood and *ichor*. This passage thus indicates a certain stature of the fluid mixture within the blood vessels, *phlebes*, of the body. In its extreme form, it confers the attribute of divinity. What could confer higher excellence (*arête* => *agathos/kleos*)?[9] Thus, the fluid within our blood vessels can confer or contribute to virtue. But what does this mean? Is the mere possession of blood a sufficient condition to produce virtue? The operation of such a process in this context introduces a worldview in which the divine and the material come together side by side to offer sometimes complementary and sometimes redundant explanations.[10] If this is correct (and there are some narrow conditions here),[11] then natural accounts of blood and its functioning will be conditioned by both.[12] Perhaps the red fluid that flows forth from war wounds is a natural entity (call this *phusis*$_1$).[13] One way to understand blood is via its material efficacy as nourishment only. But in very early contexts (pre-500 BCE) to say that biomedical events were produced materially also invoked chance (a sort of magical serendipity devoid of divinity)—later (from Aristotle and the Hippocratic Writers onward), this becomes the position of a material/mechanistic view of nature's material operation. But here *phusis*$_1$ is a complicated combination that features material properties but does not exclude mysterious, magical elements either via the agency of chance.

Another way to understand blood is via its added powers when *ichor* is added to the mix. This new mixture invokes another sense of nature: a divine, super-nature (call this *phusis*$_2$).[14] Super-nature is aligned with magic and the divine. Because it is inscrutable, there can be no external examination of it save in the context of a religious setting. Some writers on ancient science and magic such as Dodds, Hankinson, Thivel, Lloyd, Carastro, Gordon, Collins, Nutton, and Edelstein have catalogued numerous instances of magic and the divine (though in different accounts)—particularly in the context of the Hippocratic Writers (our historical terminus in this chapter).[15]

It is my contention that the philosopher of science has three options when confronting an event that does not have a clear material cause: (1) to say that it was produced materially though featuring chance (a sort of magical serendipity devoid of divinity), *phusis*$_1$,[16] or (2) that it was directly caused by a divinity, *phusis*$_2$[17] (sometimes referred to as magic in a direct sense), or (3) that it was caused by unexplainable factors (also sometimes called magic) that are linked only metaphorically to divinity or metaphorically to a material cause, *phusis*$_3$.[18] The sense of magic in *phusis*$_1$ and *phusis*$_3$ are similar. In each case there is no direct tie to divinity as there is in *phusis*$_2$, but there is a paucity of information. In *phusis*$_1$ this lack of material information logically pushes the practitioners to asserting another causal entity: chance. Now chance, as is well known throughout human history, often has a flirtatious relationship with magic.[19]

With *phusis*$_3$ the situation is not the presence of "chance," but rather the lack of critical information so that although it seems clear that there are forces at work, their exact identity is unknown. Since explanatory reason abhors a vacuum, in steps magic (metaphorically) to fill the void.

These three cases represent a common response to inexplicable natural events. But these responses need further conceptual analysis.

Underlying these three responses are worldviews that situate epistemology and metaphysics as being of a certain character. In the first instance *phusis* is understood materially. Now this is rather broad—especially in the pre-Hippocratic Writers. Often, one sort of material was given preference—such as 'air, *aer*' which becomes *pneuma* (an active vital force that can be mixed with blood or possess its own system of transmission).[20] However, if we agree that one important component in espousing natural philosophy is to present the way causation works materially, then the movement towards greater first-order input at this level is crucial. By "first-order" what is meant is an account at the level of actual occurrences.[21] For example, if someone gets a fever and there is a suggested diagnosis, prognosis, and treatment in terms of bodily fluids being out of balance (in the context of explanatory theory) and that diet intervention is necessary, then we have an instance of first-order material interventions against a materially caused condition to effect a material outcome (the cure). This is *phusis*$_1$. We are in the realm of explaining a material explanandum with a material explanans.

In *phusis*$_2$ Divinity is set out as the principal cause. Theoretically, appeal is made generally to the creative power of god(s)—or to specific gods such as Asclepius.[22] In the context of metaphysical realism, god(s) *actually* intervenes in the earthly realm concerning the causes and outcomes of disease. Though god(s) is inscrutable in this activity, the understanding is that god(s) is really a causal agent—not just generally in the creation of everything, but in *this particular fever* that Antipater is suffering. Petitionary prayer and interaction with priests in an *asklepieion* is what will bring about a cure.[23] We are in the realm of explaining a material explanandum with a real, non-material explanans.

In *phusis*$_3$ we enter the anti-real. Epistemological anti-realism refers to a position in which statements about how we know things about the world are situated within a radical skepticism about the ability to know *anything* with certainty because the objects of knowledge are either not stable or are unattainable.[24] In epistemological anti-realism, one moves away from assumed *actual* causes (be they material or nonmaterial) to linguistic devices of "*als-ob*" or heuristic discourse.[25] In the history of the philosophy of science, *heuristic reasoning* may or may not have any real connection to the world. Some very prominent philosophers have asserted that causation, for example, is not *necessary* (here a code word for *real*), but rather is a rule of thumb to guide conduct in the world.[26] Science in this context becomes a reinforcement of already held worldviews.[27] Antecedent 'x' occurs and then consequent 'y' regularly occurs in constant conjunction and an explanans is provided that does not appeal to either of the two previous senses of *phusis*. If it works well enough for day-to-day life, then what is the problem? This is one beginning of magic.

Now 'magic' is a word that means a lot of different things to different people.[28] In the ancient world it could range between constituting esoteric lore (as per Plutarch) to bogus religious explanations for natural events (as per Pliny).[29] In these two cases the writer ascribes 'magic' to an account that is different from the proponent's view of *phusis*. Thus, if one were a materialist, then any other sort of causal account is, by my account, 'magic.'

The situation for those who hold *phusis*$_2$ is rather different. Because they believe that the world is created by god(s) according to sound principles,[30] these adherents will accept material and nonmaterial explanations. The only sticking point will be the anti-realists. Since the anti-realists are radical skeptics, their account of causation is (according to the *phusis*$_2$ adherents) magical—due to there being no necessary connection between cause and effect.[31]

To make matters simpler for this philosophical study, let us set down that because I am depicting *phusis*$_3$ as anti-real, and because epistemological anti-realism asserts that we have no actual epistemic contact with real object-interactions (either in the natural or supernatural world), then any explanations set out from this standpoint will carry an element of surprise. This surprise I will also characterize as 'magic.' The surprise can be lessened when many instances yield the same results. But because of the conceptual and linguistic form of anti-realism, there could never be the sort of certainty that science demands (here understood as the Aristotelian *A. Pr.* and *A. Po. episteme*). What we are left with is some sort of probable relation between antecedent and consequence without providing a realistic causal mechanism of sufficient specification. For example, one might suggest that Antipater has a fever because he is *possibly* materially or spiritually out of balance. Both cases would be an example of *phusis*$_3$ because there is no specification of *what* material or spiritual entity is out of balance and *why* being out of balance would make a difference.

Most often the acceptance of epistemological anti-realism would entail a perversion of either *phusis*$_1$ or *phusis*$_2$. In the first instance the previous example of Antipater being generally "out of balance" without specifying the reason for this would be a perversion of a material account (*phusis*$_1$). On the other hand, if one were Euthyphro and wanted to get rid of his father so that he might inherit the *oikos* by appealing to what is "pleasing to the gods" in a circular argument, then one would have an example of perversion of *phusis*$_2$.

From the perspective of *phusis*$_1$ and from *phusis*$_2$ the realm of magical explanation is the epistemologically anti-real. It makes a nod to real understandings of nature as being properly explained materially, theologically, or a combination of the two. What sets this mode apart is its epistemological anti-realism. Necessity is out the door. What we are left with is to confront individual instances with vague, possibly relevant background conditions of the physician's personal experience in the world.[32] This standpoint in the ancient world gave rise to the medical school of the Empiricists.[33]

Of course, the worldview of the person making the judgment within the anti-real standpoint is important. For example, if Anaxagoras were an atheist, then he might declare that all *phusis*$_2$ explanations were really *phusis*$_3$ (or possibly *phusis*$_1$) because Anaxagoras denied the existence of the gods (as conventionally set out).[34] Thus, for Anaxagoras (so depicted to explore a point), all *phusis*$_2$ accounts should really be *phusis*$_3$ accounts because the reality of *phusis*$_2$ is being denied. Thus, the advocate of anti-realism would characterize religious accounts of nature and medicine as magic (in a derogatory sense).

Likewise, someone who opposed the four elements of Empedocles, for example, might relegate him to *phusis*$_3$ because he felt Empedocles' theoretical, material framework was overly fanciful and not properly supported. This judgment could come from *phusis*$_1$ or from *phusis*$_2$ advocates. However, because of the material nature of the *phusis*$_1$ claims, it is less likely that it would be termed 'magic.' Instead, it would simply be characterized as *wrong* (more on Empedocles later).

Thus, this chapter will contend that 'magic' be understood within the context of *phusis*$_2$ or *phusis*$_3$ according to the context of the critic. This means that magic defines the realm of the epistemological supernatural and the anti-real. However, most anti-realists will demur, saying that they are really in favor of material accounts, but that their understanding of *skepsis* requires them to consider many alternatives at the same time without dogmatically asserting *one* perfect account. Therefore, though both the anti-realists and the supernatural advocates are philosophically committed to the title of magic, it is the supernatural crowd who will most often be tagged as such. This dynamic will be crucial as we look at treatments of blood in the ancient world and what they tell us about scientific explanation.[35]

If it is agreed that 'magic' is generally viewed as an inscrutable causal agent, then, in one way, magic *supports* the advancement of science.[36] This is because the observer is not content to say that there is *no* causal account. ('Causal' here means some antecedent condition or conditions that consistently are responsible for a subsequent event happening along the lines of Leibniz's principle of sufficient reason.) Instead, the causal accounts are couched within (1) a material context, (2) a divine context, and (3) epistemological anti-realism.[37] This book will concentrate most on the first and third senses of *phusis* because they uniquely separate science from fiction and narrative.

These two forms of natural explanation connect to other senses of *phusis* that bifurcate natural philosophy: more concrete cases of individual natures that are tied to material non-necessary accounts, *phusis*$_3$, and more general, lawlike accounts are tied to either the divine, whose nature, beyond the anthropocentric, is conceptually more abstract or to abstract natural laws, *phusis*$_{2\text{ or }1}$.[38] Magic is introduced when there is no good nomological material account. If magic is introduced via an account that describes the actual input and interaction of the divine/magical with the material/natural, then we have an instance of *phusis*$_2$, and the attribution is directly connected to the divine (*phusis*$_2$ earlier). If it is introduced via *phusis*$_3$ earlier, then magic and the divine become logical placeholders that satisfy the stipulation that all events have antecedent conditions that brought them about even when the material understanding belies a constructed *phusis*$_2$.[39] In such cases it is more important to hold onto the maxim α that "all events have antecedent causes" than maxim β: "unless I can empirically validate a material cause, I will not commit to there being a cause." The former maxim α is a rather general worldview commitment to the way natural events operate: they all have causes. The latter β is a form of naïve Empiricism: there is no cause at work unless I can empirically pinpoint a material connection that lies right before me, *now*. Unless the data before me reveal how they came about, I cannot assert there is a cause to the witnessed event. This is a foundational posit of epistemological anti-realism and leads to a form of skepticism (see Chapter 4 this volume).

In the ancient world—particularly in medicine—a commitment to α moves one toward what would later be called *Methodism* or *Dogmatism*,[40] while a commitment to β often leads to a form of *Skepticism* or *Empiricism*. The modern (post-Kant's first *Critique*) understanding of the philosophy of science suggests that α is a more fruitful conjecture for the development of science than β. The problem is that when one is dealing in nascent science, commitment to β can get in the way of affirming α. This is because there are so many events that appear to have no discernable cause. The most logical response (from the *phusis*$_3$ standpoint) to such a database is to say that many to most events are random and that natural causation (an essential posit to the *phusis*$_1$ worldview) is sporadic and not "lawlike."

This is a lot to lose.

As suggested by Kant, it may be a better strategy for those who aspire to explanations based on *phusis*$_1$ to affirm α even if it means the acceptance of *phusis*$_2$ (albeit in the constructed form).[41] To some extent, many physicians today do this when faced with a patient who has a disease that is not very well understood or for which there is no cure. The physician, in many cases, will make reference to possible cures or remissions on the basis of magic (unsubstantiated experimental treatments) or divine intervention (aka prayer to god).[42] There are cases in which people with cancer, for example, go into remission without a clear material account that can be articulated. Some attribute this to *phusis*$_{2 \text{ or } 3}$ (either literal or constructed).[43] Because our *phusis*$_1$ accounts are much more extensive than they were in the time of the pre-Hippocratic Writers, recourse to this secondary account is much rarer. But the dual explanation account still is remnant—even in a vestigial form.

The aforementioned philosophy of science speculations are based upon logical consistency. However, in practice, it was often the case that all of these senses of *phusis* were practiced together without a sensibility of competition. This was especially true in the sixth and early fifth centuries. As Vivian Nutton notes, "Healers could act like Empedocles as roving shamans, and the boundaries between magic and medicine were almost non-existent."[44] Eric Robertson Dodds agrees, as he suggested "a tentative line of spiritual descent which starts in Scythia, crosses the Hellespont into Asiatic Greece, is perhaps combined with some remnants of Minoan tradition surviving in Crete, emigrates to the far West with Pythagoras, and has its last outstanding representative in the Sicilian Empedocles."[45] Spiritualism and materialism existed side by side. One could go to an *asklepieion* and engage in the incubus therapy of sleeping with nonpoisonous snakes in the night and then having your dreams interpreted the next morning by a priest and then consulting someone else who might prescribe organic herbs.[46] Or as Edelstein conjectured, during this early period, those who created a reputation for *phusis*$_1$ were elevated to *phusis*$_2$ as a sign of reputation recognition.[47] They were recognized either as being godlike or as having a close connection to a god of healing.

Returning to the heart and blood in Homer, these two sorts of explanations interact. In a passage in the *Iliad*, the heart is related to courage:

> 'You wine sack, with a dog's eyes, with a deer's heart. Never
> Once have you taken courage in your heart to arm with your people
> For battle, or go into ambuscade with the best of the Achaians.
> (I. 224–226, Lattimore, tr.)

The sense of this text is that Atreides could have acted otherwise. He is being berated for not putting his heart into a certain condition in which it and its blood are courageous. Rüsche read such passages as indicating that emotion, *thumos*, resides in blood.[48] Now to round out this reading requires some unsubstantiated conjectures. Aristotle associated the heart with the production of blood.[49] If one accepts this posit, then if an individual acts in a certain way via the agency of blood and his heart, then surely everything associated with that process will possess the same property. Thus, if courage is to be gained in the heart, then the blood produced in the heart must also possess this property.[50]

One possible problem with this conjecture (from a material causal stance) is the possibility that the heart is also the source of thinking. Aristotle, for one, believed this to be the case. Now, if the Homeric Writers (writing well before Aristotle) held the view that cognition is in the region of the heart (as physically situated or dualistically concurrent), then it may say nothing about how courage arose in the physical heart (cf. *Iliad* XXI. 441, *Odyssey* IV. 572, V. 389).

This is where the earlier passage on *ichor* is of value. We can return to the two standpoints. *Ichor* may just be a device of magic linked to the divine. If the gods have *ichor* in their body and if humans have *ichor* mixed with blood in theirs, then there is a vehicle established for the expression of magic. On the other hand, if one were to view *ichor* materially, then if the heart and blood were such that there is a propensity toward courage (that could or could not be realized), then blood takes on the role of what we in the modern world might analogously call a "genetic predisposition."[51]

Either reading (materialism or magic) might connect to resonant themes of fate that are very important in both the *Iliad* and the *Odyssey*.[52] At any rate, some sense of personal freedom seems to be clear here such that either one of several scenarios pertains: (1) one's blood has a connecting factor with the gods, *ichor*, that allows magical demonstrations of virtues to come forth; (2) one's blood has a particular character (perhaps from ancestors) that materially (or via ancestor spirits) determines individual choice; (3) one's blood has a particular material character that dynamically interacts with one's mind in the process of decision making; or (4) a person plainly chooses and this act determines the character of his heart and blood (as a consequence of this choice). From the text, any of the four readings is equally speculative. This ambiguity is played out in *Iliad* X. 220:

> 'Nestor, my own heart and my own proud spirit arouse me
> To go into the host of the hateful men who lie near us.

Now, this could mean either that the heart and its by-product, blood, materially *cause* Nestor to act, *or* it might mean that there is a dynamic interaction between the mind and the heart, *or* it might mean that the mind (in the region of the heart) acted of its own accord. The text is silent on this. It is the conjecture of this treatise that the blood/heart possess some material properties of their own that interact with the mind—wherever it is located (in the region of the heart or brain).[53]

Another curious point about the *Iliad* 5.339–442 passage is that it suggests that normal blood is created as an end product of eating. One eats food and it is processed into nourishment, *trophe*, and this is delivered to the body in the form of blood.[54] Thus, blood is a natural by-product of a bodily process. The component of *ichor* is separate from this process (at least in the case of the gods). This magical, godlike property of *ichor* is repeated at 5.416. This is thus an instance of mixing magic and material accounts.[55]

As suggested earlier, *ichor* is not reserved for the gods. In Aeschylus' *Agamemnon, haima* and *ichor* are conflated in two lines with reference to a mortal wound:

> From him deep in the nerve is given
> The love and the blood [*haima*] drunk, that before
> The old wound dries; it bleeds [*ichor*] again.[56]

This conflation is continued in the Hippocratic Corpus until the writings of Aristotle.[57] But even here, the fundamental quality of mysterious life-giving power is not lost.[58] The Hippocratic Writer declares in the work *On Wounds to the Head* that in the case of a fatal wound to the skull, "the cut becomes discolored as a bit of the *ichor* flows from it."[59] The point here is that *ichor* is associated with the loss of life itself. We can conclude from these passages that the ancients in the Western tradition were keen on maintaining both a material, physical understanding of blood as arising from food and providing nourishment, but also another mysterious, quasi-magical sense of the blood through *ichor* in which nonexplainable operations of the body—such as thinking, the passions, and virtue—might be given an account that is in part physical and in part magical. The strategy here was to identify causes. One causal account might be that blood (*haima*) is physical and if the physical constituents are faulty (from the food that is used in its construction, for example), then physically generated consequences follow. Another quasi-cause might be that blood (*ichor*) is mysterious and has properties that are hard to explain except to attribute them to this largely inscrutable causal origin.

Life is so mysterious. How is it that we are alive one day and die the next? There may be many suppositions about this (derived by context), but one powerful possibility is that there is in our blood a combination of aspects of material explanation with nonmaterial explanations. Such a mixed account is resonant with other early writers on the subject.

The first Western natural philosopher who made inroads on this question is Alcmaeon of Croton. The dating of Alcmaeon is very conjectural. In Aristotle's *Metaphysics* 986a 22, it says that "Alcmaeon was a young man in Pythagoras' old age." Under this tradition (supported by Diogenes Laertius),[60] Alcmaeon was said to have been a pupil of Pythagoras. This would put him roughly as a contemporary of Empedocles. But Ross rejected the *Metaphysics* quotation as being corrupt. This fits into the dating by Edelstein as early fifth century.[61] However, Guthrie[62] and Harris[63] argue that there is no clear proof that Alcmaeon was *not* a student of Pythagoras so that if Pythagoras died around 490 BC, then Alcmaeon could have been born 510–515 (Empedocles' birth is often put around 500.) At any rate, this chapter will place Alcmaeon as a little older than Empedocles.

As mentioned earlier, the ancient Western world engaged in a controversy between the seat of intelligence: whether it was in or around the heart or in the brain. There is no clear evidence for the originator of the brain as the seat, but something can be made of the fragment from Philolaus of Tarentum (a probable follower of Alcmaeon and Pythagoras, circa 470–385)[64] who said that

> there are four vital organs in the rational, living being: the brain, the heart, the navel, and the genital organs. The brain is the seat of the mind, the heart of the soul . . . and of sensation . . . the navel of the plant, and the genital organs of all of them.[65]

This passage is interesting in several ways. First, it makes a separation between intelligence and the life force. Second is the assertion of a link to other life forces—here the plant. This is reminiscent of Aristotle's three-soul theory—plant (nutrition), animal (sensation, locomotion), and man (reason)—cf. Rüsche on *psuche* in the blood.[66] Finally, the fact that the genitals contain them all situates him within the emerging pangenesis debate that would sharpen in Hippocrates and Aristotle.[67]

Another key issue that is brought to the fore by Alcmaeon is the mixture of air in the blood. Now *air* is understood variously by the pre-Socratics. Air according to Anaximenes was the original substance of the universe that changes according to condensation and rarefaction (Aristotle, *Metaphysics* 984a 5). A second sense of air is when it is an active vital element (often when heated). In this case, it becomes *pneuma*. *Pneuma* is a vital force in the body that is generally conveyed by blood (perhaps supplanting the earlier *ichor* as the $phusis_2$ account). Theophrastus (DK fragment 212) and Aëtius (DK fragment 214) seem to suggest an attribution to Alcmaeon of the doctrine that air (as *pneuma* or converted into *pneuma*) was breathed directly from the nose to the brain. This transmission method would quickly alter the mixture of blood in the brain and give it an immediate source of *pneuma* (perhaps explaining why the brain, according to Alcmaeon, is the most powerful organ and the seat of reason and assuming that *reason* is the most powerful capacity and that *pneuma* is the most powerful building block). Since *pneuma* is so powerful and so plastic in its expression, another way to conceptualize it is in the modern example of stem cells. Stem cells are vital and pluripotent so that they can adapt to their environment and repair organs and tissue that are not functioning properly.[68] The analogy is very loose, but it gives the modern reader a sense of how pluripotent and plastic *pneuma* was thought to be.

Harris mentions a possible distinction that Alcmaeon may have made between *phlebes* with blood and those without blood.[69] Sleep is said to occur when blood retreats partway out of the *phlebes*, and death occurs when the retreat is complete.[70] Whether this describes one process or whether it describes a distinction between arteries and veins is unclear. But this author would side with the process account and leave the discovery of the difference between arteries and veins to others, for example, Praxagoras and Erasistratus. Thus, for all these authors, *phlebes* will refer both to veins and arteries.

Finally, if we can rely upon the supposed connection of Alcmaeon to Pythagoras and if we can rely upon Diogenes Laertius, then another doctrine relevant to our inquiry can be set out, and that is an extension of the aspects of intelligence and soul alluded to earlier. Diogenes says,

> The soul of man, he [Pythagoras] says is divided into three parts, intelligence [*nous*], reason [*phrenas*], and passion [*thumos*]. Intelligence and passion are possessed by other animals as well, but reason by man alone. The seat of the soul extends from the heart to the brain; the part of which in the heart is passion, while the parts in the brain are reason and intelligence. The senses are distillations of these. Reason is immortal; all else is mortal.[71]

This quotation speaks to the ancient controversy on whether the brain or the heart (or the blood around the heart) constituted the seat of reason. This is an important question because it is precisely *this* that would separate man from beast. The quotation suggests that both the heart and the brain were seats of consciousness (animal/human) and that emotions exist solely in the heart. It is beyond this treatise to fully explore whether this doctrine can properly be attributed to Alcmaeon, but if we are more interested in the development of ideas rather than precise attributions, then this combination sets the stage of the later debate between the so-called Cos Hippocratic Writers (pro brain for reason) and the so-called Sicilian writers/Cnidian Hippocratic Writers, and Aristotle (pro heart for reason).[72] However, Theophrastus does say that Alcmaeon was the first to separate sense perception and consciousness (common to animals and humans) from intelligence (the province of humans alone).[73]

The last major writer to be examined is Empedocles. Empedocles was keenly interested in the linkage between the four roots—shining Zeus, life-bringing Hera, Aidoneus (Hades-Death), and Nestis: fire, air, earth, and water (DK fragment 6)—and blood. The statement of these four roots sets out all three forms of *phusis*, though it emphasizes the first and second senses more. The reference to the gods is an expression of *phusis*$_2$. Since the four roots are fundamental principles upon which one might construct a wholly material account of nature (*phusis*$_1$), any connection to these roots may be classified as a primary broad explanatory principle that goes beyond the empirical evidence (*phusis*$_3$). In this way, Empedocles unites the three perspectives of scientific explanation.

There is another way in which the four-element theory makes for a more complex account of blood and other natural things. This is because unlike some of the previous writers who posited only one ontological entity (such as water, air, or the unlimited), a pluralistic account that involves mixing lays the groundwork for the four-humor theory in the Cos Hippocratic School. It also lays the groundwork for Aristotle, who combines these further with the contraries: hot, cold, wet, and dry. This combination of properties works on the level of *phusis*$_1$.

Blood as *phusis*$_1$ comes about through the activity associated with the next level of expression, *physis*$_2$, in DK fragment 98:

> And earth chanced in about equal quantity upon these, Hephaestus [i.e., fire], rain and gleaming air [*lit.* aither], anchored in the perfect harbours of Cypris [i.e., love], either a little more of it or less of it among more of them. From these arose blood and the various forms of flesh.[74]

There are a couple of important points to be gleaned from this quotation. First is that blood is slated to be at the next level of construction from the elements themselves. Second is that blood is compared to flesh. This coincides with the later distinction by Aristotle of the *homeomerous* parts (those parts that are uniform in themselves) and the *anomeomerous* parts, such as an organ like the heart that is varied because of a functional architecture.

One of the controversies in the ancient world is in the more intimate connection between flesh and blood via skin respiration—often attributed to Empedocles. In DK fragment 100 from Aristotle's *On Respiration* (473a 15), there is the famous reference to a girl playing with a bronze vessel called a clepsydra in a pool of water. The girl puts her hand over the narrow opening of the otherwise closed vessel and submerges it into water and then lets her hand go. The air rushes out and the water rushes in.[75] Clearly, this action is seen to be driven by material forces (*phusis*$_1$). Just as the clepsydra was a vessel with a small opening at the tip, so also might there be small pores in the skin through which air rushes in and interacts with blood (taken to have similar properties as water). The air rushes into the pores on the skin and mixes with the blood under the skin. If this is the correct reading of this passage (though some think that in the passage the pores refer to the lungs), then there are three forms of respiration that have been set out: (1) through the nose directly to the brain (Alcmaeon), (2) conventional lung breathing via the mouth and windpipe, and (3) skin breathing via the pores in the skin (Empedocles). All of these accounts of respiration are different ways in which blood can receive air and transform it into *pneuma* (though the process here seems to be a fuzzy application of *phusis*$_2$). If we accept the suggestion that Philistion was a successor to Empedocles in his school and that he endorsed the theory of respiration set out in Plato's *Timaeus* 79a, then the air that is materially transformed into *pneuma* is responsible for keeping the blood in constant motion (*phusis*$_1$).[76] Therefore, we can surmise from this data that direct nose respiration is necessary for the brain's added need for vital *pneuma*. The lung source of respiration might serve most of the needs of the body, but has to be supplemented by skin/pore respiration. This is highly speculative, but is set out in the spirit of *phusis*$_1$.

The primacy of blood is emphasized again in DK fragment 98 and KR fragments 392 and 394. Empedocles says that it is via the agency of blood (*haima*) that we think:

> . . . So especially with the blood that they think; for in the blood above all other parts that the elements are blended.[77]
>
> . . . [The heart] dwelling in the sea of blood which surges back and forth,[78] where especially is what is called thoughts by men; for the blood around men's hearts is their thought.[79]

In these passages, the sentient properties of the blood are highlighted. As mentioned earlier, the location in the body where reason occurs was a disputed question in the ancient world. The two principal candidates are the brain and heart (or area around the heart), but here it seems that Empedocles is more interested in blood in the context of a vital principle of soul (*pneuma*$_3$).[80]

Clearly, in the ancient world (and today), the causal account of thought is a very difficult concept to understand. There obviously is a material ground to thinking (*phusis*$_1$ as in our modern neuroscience brain models), but the precise mechanics are still out of reach. Either this is because our science in the modern world is too primitive or it is because there are complicated levels of systemic interaction in which physically generated processes are not best explained at the lowest level of physical organization (the so-called reductionistic hypothesis, *phusis*$_1$ versus the emergent theories hypothesis, *phusis*$_{2 \text{ or } 3}$).[81] In the ancient world, this reductionist point of view would be evinced by the atomists. When one begins with only atoms, the void, falling, and *prolepsis*, then the account is firmly directed toward *physis*$_1$. The problem in science, then as now, is that material accounts are incomplete. We respond to this by saying (1) we don't have the answer, but we believe either there is an unknown, definitive answer or it is a chance event (the etiology of some genetically based nonheritable diseases is like this);[82] (2) it is actually caused by divine intervention (either through prayer or a miracle)—recoveries from illness that seem totally contrary to modern science are like this; or (c) we attribute it to divine intervention or to materialism only metaphorically (meaning that it is an epistemological problem and not a metaphysical event). Thus, even today we respond analogously to a similar problem with the same philosophical approach as the pre-Hippocratics, viz., the layering on of various senses of nature.

In the ancient world, one way to account for material principles (such as health and vitality) and magical principles (such as life and thought) was to create two carriers: carrier one, *haima*—blood as the transporter of ordinary material properties, born of food and delivering nourishment to the body—and carrier two, *ichor* and/or *pneuma*—the bearers of thought, life, and possibly reproduction.

Thus, blood and its accomplices (the heart, the blood vessels) have a physical *phusis*$_1$ part (associated with *haima*) and a magical *phusis*$_2$ or $_3$ part (associated with *ichor, pneuma*, or *tuche*). Harris cites Diogenes as creating a theory in which *pneuma* coexists with blood in the blood vessels, saying that *pneuma* in its rarest forms is a "supremely spirituous form as God . . . the doctrine which will identify *pneuma* in the Stoic teaching with the spark of the world-soul."[83] This is yet another instance of magic.

Now if magic is taken as a given in life,[84] then both accounts might seem plausible to hard-headed speculative thinkers of the time. They both came from different sources, but neither had a firmer claim on truth. It has been this author's contention that the ancient Greeks were not afflicted with the West's post-Ockham obsession with simplicity.[85] Redundant explanation was just fine. This was because nature, *phusis*, was so complicated that more than one account was welcomed. Aristotle's four causes for a single given event is a prime example.

If this acceptance of redundant explanation is correct, then it would not seem out of place that a complete account of blood and its accomplices might include redundant explanatory devices; for example, for the heart and blood vessels, one would point to the material properties of blood *along with* its magical properties.

> Principal among the magical properties is reproduction. In *Iliad* 19.111:
> ". . . Come then, lord of Olympus, and swear before me a strong oath
> That he shall be lord over all those dwelling about him
> Who this day shall fall between the feet of a woman
> That man who is born of the blood of your generation.". . .

The reference to the blood of a generation indicates that blood is connected with reproduction. The magical side of blood is first associated with life itself (*phusis*$_2$ and $_3$)—both its mysterious creation and destruction. The material side of blood is borne through the female menstrual blood, *katamenia*, thought to be the female seed (in the dual seed theory), and the male *sperma* that comes about via a concoction of blood in the testes.[86]

There is a strong tradition of this, which develops through Hippocrates, Aristotle, and Galen.[87] Though there is the outline of a *phusis*$_1$ account, it is not complete enough to constitute an explanation. In the Hippocratic Writers, writing on the tradition that preceded them, there is call for a more exact treatment:

> I say not that man is air, fire, water, or earth [the Empedoclean roots]—or anything else for that matter . . . these writers do not give an account of 'what is' (*ti esti*).[88]

As mentioned earlier, the pre-Socratic writers, such as Thales, Anaximander, and Anaximenes, had posited a single underlying principle for everything: water, air, or the unlimited.[89] Anaxagoras and the atomists offered a multiplicity of elements through the seeds and atoms. In this way, Empedocles seemed rather in the middle. As a medical practitioner himself, he saw the benefit of the 'empirically concrete' in accounting for nature. He also saw that having a small, core number of principles was better than a multiplicity. The Hippocratic Writers for the most part are not necessarily against this (though in different ways). But what they are insisting on is that the roots be used to construct even more concrete fluids that actually exist in the body. In this way, the Hippocratic Coan Writers seek a more systematic, philosophical approach.[90] With respect to blood, this means putting it into context with other primary substances (such as the four roots as well as the contraries: hot, cold, wet, and dry)—a task left to Aristotle.

The unique physical properties create a unique variety of effects. What is of interest here is that these paired opposites imply a tension of extremes and the possibility of median balance. Alcmaeon builds upon the mathematical model of balance developed by Protagoras and applies it to health in his (Alcemaeon's) doctrine of *isonomia*.[91] The goal is to have them all balanced so that there is not too much or too little (DK fragment 4). When this balance is lost, pain and ill health result. The physician's job, then, is to restore balance. This process requires first an assessment of which humor is out of balance (a mathematical inclination). If there is too little, a targeted stimulant for that principle out of balance (cf. later humor theory) is in order. If there is too much, then a strategy to draw down the imbalance is necessary. In the case of blood (as per our argument), there is the material part and the magical part. In the material part, there is the soul (*psuche*$_1$—following *phusis*$_1$) of everyday life (6.10). Though the author does not directly say so, it seems reasonable to associate this with *thumos*, or spiritedness. When one has too little blood, then he is rather lacking enthusiasm. Too much blood results in overly high spirits or compulsions. Thus, Alcmaeon forms the intellectual groundwork for the subsequent Hippocratic theory of health as balance among opposites that is captured later in the theory of the four humors.

The primacy of blood as one of the principle materials of human concern reflects the Homeric origins of the battlefield. This is set out later in Hippocratic texts such as *Wounds in the Head*—the loss of too much blood is seen as a proximate cause of death. Since the beginning and ending of life are associated with blood, and blood is associated with *psuche*$_2$, then this second sense of soul is also associated with the beginning and ending of life (inexplicable, aka magical events—*phusis*$_2$). Geoffrey Lloyd has documented this connection to the magical, citing:[92]

> *On Regimen* I 10 [Hippocrates] has a doctrine of three 'circuits' in the body corresponding to the three circuits of the heavenly bodies.

Clearly, the link to the heavenly bodies is a stretch if one is from the material nature side *phusis*$_1$ standpoint. But from the magical *phusis*$_2$ or $_3$ standpoints, this is very plausible—especially since the link to the heavenly bodies has a hard-nosed mathematical flavor.[93]

In the *phusis*$_1$ standpoint, the connection of blood to its attendant structures, the heart and the blood vessels, is important. Aristotle cites three writers who preceded him as discussing the vascular system: Syennesis of Cyprus, Diogenes of Appollonia, and Polybus:[94]

> *Syennesis*: The nature of the stout blood-vessels is this:[95] from the eye across the eyebrow, along the back, past the lung, below the breasts; one runs from right to left, the other from left to right: the latter runs through the liver to the kidney and the testicle, the former to the spleen and kidney and testicle, and thence to the penis.
>
> *Diogenes*: The blood-vessels in man are as follows:[96] There are two which are outstanding in size and these extend through the belly along the backbone, one to the right, the other to the left, each to the leg on its own side of the body, and upwards to the head past the collar bones through the throat. From these two, blood-vessels extend all over the body: from the right-hand blood-vessel to the right side, and from the left-hand blood-vessel to the left side; of which the most important are two which run to the heart near the backbone, and two others which run slightly higher up through the chest under the armpits, each to the hand on its own side of the body: one of these is called the splenetic, the other the hepatic. Each of them splits into two at its extremity, one branch going to the thumb, the other to the palm of the hand; and a number of tiny blood-vessels, much branched, lead off from them to the fingers and the rest of the hand. Other blood-vessels, tinier still, lead off from the principal ones. . .
>
> *Polybus*: There are four pairs of blood-vessels.[97] One pair runs from the back of the head through the neck on the outside past the backbone on either side till it reaches the loins and so to the legs, and after that through the shins to the outer part of the ankles and to the feet. That is why surgeons make incisions in the hams and outer parts of the ankles to relieve pains in the back and loin. Another pair of blood-vessels runs from the head past the ears, through the neck: these are called the jugular veins. They continue inside along the backbone, past the loin-muscles to the testicles and to the thighs, and through the inner part of the hams and through the shins to the inside of the ankles and to the feet; and that is why surgeons make incisions in the hams and [inner sides] of the ankles for pains in the loin-muscles and testicles. . .

These texts are interesting because they show investigators trying to give *phusis*$_1$ accounts of the vascular system and its relation to clinical practice. It is *because* of the fluids within and the direction of the blood vessels in the body that certain "surgical" interventions might be made.[98] In each passage there is theoretical bias for bilateral symmetry.[99] Whether this pattern was derived inductively through observation (according to *phusis*$_1$) or whether it was imposed upon the phenomena via broad divine principles (according to *phusis*$_2$), it is impossible to say. There was in the ancient world a disposition to "save the phenomena."[100] Sometimes this required imposing patterns from mathematics or astronomy upon other natural entities in order to continue the same schema.[101] If there were an underlying theoretical connection, then such a move would be akin to theory "projection" as is practiced today.[102] However, this is rarely the case in ancient science. Lacking an underlying comprehensive account, such projections are more akin to magical or artistic patterning (*phusis*$_2$).

CONCLUSION

What we have discovered at this first stop on our voyage is blood, an essential fluid for life, has moved *from* being a qualitative, unspecified composition of vague material components (*phusis*$_1$) with *ichor* (*phusis*$_2$ and $_3$) (in the Homeric Writers) *to* a more defined material explanandum (*phusis*$_1$) that evokes various, more specific material explanans (particularly in Alcmaeon and Empedocles) along with magic and the divine (*phusis*$_2$ and $_3$) brought in only when the former cannot account for the phenomena (the atomists stand as an exception to this trend, preferring the inscrutable *prolepsis—phusis*$_3$). Then, in the spirit of saving the phenomena and giving some sort of account (according to something analogous to Leibniz's Principle of Sufficient Reason), a magical explanation is produced. However, it should be noted that in these cases there is a mixture of material and nonmaterial explanation (viz., magic and the divine either intended actually or metaphorically). The amount of the nonmaterial explanation in the Homeric passages is definitely greater than in Alcmaeon and Empedocles (and their contemporaries). However, it should be noted that nature is understood equivocally as materially caused and nonmaterially caused. This results in occasionally redundant explanations, but references to *phusis*$_2$ and $_3$ are intended to fill in the gaps.

The strategy of ancient scientific explanation can be compared very generally to the way we create explanations today in which we also try to provide material accounts of biomedical phenomena, but when we are stumped, we lean on the nonmaterial because all of us *abhor an explanatory vacuum*. The physician may confront his terminal patient and say that medical science does not possess the cure for her disorder. Medical science may not very well understand the disorder. Therefore, if the patient believes in the divine, it becomes the time for prayer and hope.

It has been the contention of this chapter that the understanding of 'blood' during this era in early Greek history is a window upon a clear transition between uncritical religious/magical theories and an emerging scientific/material understanding. Because of its very nature, blood via its material components and its magical components constituted the perfect vehicle to move natural philosophers to the next level of critical inquiry. It thus affords us a preview of the ongoing journey that is the development of biomedical science.

NOTES

1. Aristotle will only be used to reinforce points made by earlier authors, see Boylan (1982, 1983, 1984, and 1986a). More will be said about the Hippocratic Corpus in the beginning of Chapter 2, this volume, along with the status of Anonymus Londinensis.
2. Though Herodotus put Homer as being around 400 years earlier (*Herodotus* 2.53); thus around 850 BC, I will date him on the later side of the general agreement of 7th to 8th century. This later dating will make the quotation from the *Iliad* more proximate to the pre-Socratic philosophers.
3. An analysis that follows this rather gruesome mode of discovery was set out almost a hundred and fifty years ago by Darenberg (1869).
4. Of course, some latter writers—especially Aristotle will diminish the mystery associated with blood. But not all writers. See Jouanna and Demont (1981).
5. Hippocrates, *On Fleshes*. 11, Plato, *Timeas* 83c, c.f., Aristotle, *H.A.* 521a 2, 521a 18; *P.A.* 651a 18, cf. Galen, Kühn 15: 345.
6. Aeschines, *Epistulae*. 3.160.
7. Adkins (1960), p.314 on the general disposition to situate morality via biology, cf. pp. 65–70 on the gods as moral, cf. Marie-Paule Duminil (1983): 164–180.
8. Lattimore, translation; cf. *ibid*., in accusative case ιχῶ, 416, in the plural, *ibid*. VI, 3–5.
9. For complementary accounts, see Adkins (1960) and Olson (1995).
10. See Lloyd (1979), Van der Eijk (1990), Hankinson (1995c, 1998a, 1998b, 2003, 2005).
11. Jouanna and Demont (1981) set a very thorough context for viewing this passage: "the blood of the gods" in the context of both *Iliad* V. 416 and *Agamemnon* V. 1480. In these latter passages there is a possible controversial transition from "the blood of the gods" to an excretion as the result of a wound. This continues to the Hippocratic writers where the wound is extended to ἕλκος. In the passage from the *Agamemnon* it reads "the old wound is dried up but once again bleeds (ἰχώρ). This is the transitionary usage. See Chapter 2 on ἰχώρ in the Hippocratic Writings in this transitionary phase. However, Jouanna and Demont persuasively argue that the primary Homeric sense extends at least through the Hippocratic Writers.
12. Diels-Kranz lists two other pre-Socratic usages of ἰχώρ· τὴν χαλὴν ἰχῶρα τῆσ σαρκός, Philolaos, A27 (from Menon, Anonymus Londinensis); αἵμα καὶ οὐχ ἰχώρ, Anaxarchus A1. These are not terribly helpful, but they do show a trend to a human discharge understanding of the word that makes it separate (in humans) from blood. In Chapter 2, this volume, the development of ἰχώρ away from "the blood of the gods" to more mundane usages will be made clear.
13. The use of *phusis* in Greek thought is the subject of legal and scientific speculation. In the former case, it is often contrasted with *nomos* as per Beardslee (1918) and Heinimann (1945). However, the link to the divine in Aelian has been argued by Gordon (1987) and by Lloyd (1987) concerning the pre-Socratics, p. 49, n. 163, and Kahn (1979). On the later writers, see Hankinson (1998a, 1998b): 259–261, 287–291. For a general survey of magic in the context of magical figurines (that facilitate the magic) see Collins (2003).
14. I will use the term 'divine' to refer to such expressions as *theos, daimon*, and *to theion*. These are obviously different, but they demonstrate various general features of explanation that are similar, so these common expressions should be the point of focus.

15. Good general accounts can be found in Lloyd (1979), Hankinson (1999a, 1999b), Thivel (1975), Carastro (2006), Gordon (1999a), Collins (2003), Nutton (1985, 2013), and Edelstein (1967).
16. Carastro (2006), Gordon (1999a, 1999b, 2002), Tambiah (1985); cf. Democritus, Leucippis, and Aristotle's discussion of spontaneous generation of insects in dung, Ar, H.A. 539a22, 551a 1–5, cf., bees in G.A. 760a. Some might demur that the atomists can only be connected to chance as unpredictable. They certainly didn't want to categorize chance as under divine control. But magic might be some middle ground between the two.
17. See Edelstein (1967): 205–246 and Lloyd (1979), ch.1.
18. This is the position of Hankinson (1998a): 20 and Thivel: 76. Aligned to the metaphorical position is that of the social construction of divinity and magic; see Carastro (2006) and Gordon (1999a, 1999b, 2002). Collins suggests a personal psychological reconstruction (2003): 18.
19. Note today the worldview of gamblers at casinos. Though they know that there is no divinity involved when they roll the dice, still they are highly superstitious in how they explain victory or defeat.
20. Nutton (forthcoming) says that air can be related to the divine and to positive and negative material accounts, mss. 2–3.
21. "Second order" would refer to speculation about the principles that govern the first order. They are thus *meta-principles*. Much of this book will be toggling between these two ways of confronting blood as an instance of natural philosophy.
22. I am in basic agreement here with Nutton (2013) on the relationship between religion and medicine in the pre-Hippocratic era: 104–115.
23. For example, see Aeschylus, *Eumenides* 62 and *Suppliant Maidens* 263, for examples of involving medical healing with petitionary prayers to the gods.
24. The epistemological debate in contemporary philosophy of science is contentious. Part of the discussion comes from post-Modernism critics of epistemological realism. For the most part, I follow the analysis of Helen E. Logino (2002) in that she acknowledges the force of the anti-realist position, but creates an empirically based counterargument. For the reader unfamiliar with this debate, I would suggest the general introduction by Stuart Brock and Edwin Mares (2007). A few point-counterpoint discussions will also clarify what is at stake. Michael Resnik (1997) provides a good example of a pure mathematical anti-realism stance that is objected to by Mark Balaguer (1999). In the medical realm of psychotherapy, see Barbara Held (1999), who is a critic of anti-realism in a clinical context. For an opposing view, see Katherine Morrison (1999). An interesting proponent of anti-realism within the context of the present "language context" can be found in Hartry Field (2009). The slant of the religious realists can be found in René Van Woudenberg (2007).
25. This is related to *sozein ta phainomena*, which was an important principle in ancient science. In this instance, if we can offer an "*als-ob*" explanation that would square the phenomena with the theoretical second-order structure, then this would be a good thing. For the most influential discussions of this, see Thomas S. Kuhn (1957, 1962). In more recent work, the literature has focused upon *prejudice-problems* and not just the *process of discovery*. For some of these debates, see William Wimsatt (2006), David Magnus (1997), and Diderik Batens (1989).
26. David Hume (1748) claims an interest in sensation, yet comes up with three remarkable conclusions that show how skepticism can lead to *phusis*$_3$. The first is: The laws of science are not necessary and are not a priori (pp. 16–20).

The second is: There is no way that science or experience can necessarily prove that the future will resemble the past (pp. 21–25). The third is: What is commonly called cause is really contingently linked objects and not properly called causation (pp. 49–53). On my account, this logically commits Hume to accept that each and every natural event will occasion surprise that it really happened. There would be no surprise were there necessity. This would mean that Hume is logically committed to magic as I have defined it. However, Hume vigorously denied such an attribution in his refutation of miracles (pp. 73–93). This is because Hume was a vigorous agnostic (he couldn't be an atheist on his own epistemological standards—see *Dialogues on Natural Religion*). On my parsing of terms, this would constitute a denial of nature as *phusis*$_2$ and nothing more.

27. The realm of the "already held worldviews" is dominated by various modes of constructing a model upon nature and using that constructed model to explain it all. Of course, this is the transcendental of Kant in the first *Critique*. But others have fashioned various other flavors in this exercise. For example, Koen Vermeir (2005) discusses the interplay of magic and the mechanical in the "magic lantern." The use of analogy in this construction gives a flavor of empiricism. The enthymeme is that the practitioner chooses the *suitable* analogical example. Another important analysis comes from Charles Webster (2005). Webster's argument that the gap between magic and science is not overly broad is in several ways consistent with my contention that magic is really the practicing of science by epistemological anti-realists. On a different track is Edward Karshren (2011), who tries to engage *phusis*$_2$ with a semantically constructed cosmology. This would, in my terms, move the evaluation of *phusis*$_2$ into the realm of *phusis*$_3$.This, of course, is a general tactic of anti-realists who eschew supernatural reality. Peter Struck (2004) sets out symbols as ontological signifiers. This is consistent with my portrayal of there being magical significance in the operation of anti-realism in science via *phusis*$_3$. I must assert that this is still a very strong question. Much of the question and answer period when I presented this chapter to a seminar at the University of Paris IV (October 2013) revolved around this question. I am still engaged with audience members about this central controversy.
28. Michal Buchowski (1988) sets out a classification scheme of five sorts of magic. The *Oxford English Dictionary* cites various meanings associated with ritual activities and natural magic. A key citation in English occurs from Chaucer, *The Squires Tale*, 218 (1395): "It is rather lyk an apparence ymaad by som magyk." This connotes the meaning of anti-realism in *physis*$_3$. What is inexplicable is accepted that way or pushed into a perversion of a material explanation or a theological explanation. This inexplicable wondering is also present in Shakespeare's *Winter's Tale* V.3.39 (1623): "Oh Royall Peace: There's Magick in thy Maiestie."
29. Plutarch, *Themistocles*, 29.5; Pliny, Bk XXX., 1:"I have often refuted the fraudulent lies of the Magi. . ."and Bk XXX, 2 "Magic arouse in Persia with Zoroaster;" and then a discussion about bogus cures for toothache, XXX, 8.
30. An obvious example is the Theogony.
31. Some readers may demur, saying that I am importing contemporary concepts upon the ancient world. I would reply: guilty as charged. This book is about a philosophical study of blood using all concepts and distinctions that have developed in the Western tradition concerning the philosophy of science.
32. Here we have the following situation: since there are no actually known objects, there can be no firm rules—only *pro tanto* accounts that are subject to the physician's personal experience in the world along with his *novel* (and

possibly *unique*) assessment of this patient before him. It all begins all over again each and every time.

33. For some background on the Empiricists, see Edelstein (1967): 195–204; Nutton (2013): ch. 13 et passim; A. A. Long (1986): ch. 3; A. A. Long and D. N. Sedley (1987): vol. 2, 1–17; 458–475; and Boylan (2004b), "Galen" *The Internet Encyclopedia of Philosophy: http://www.iep.utm.edu/galen/* [accessed July 1, 2014].
34. It is certainly unclear whether the actual Anaxagoras held such opinions. But the people of Athens believed he did and used it as one of the two charges against Socrates in the *Apology* 26 c-e. His doctrine of νοῦς DK 14—ὁ δὲ νοῦς , ὂς ἀέί ἐστι could be a stand-in for *phusis*$_2$ while the ʽσπέρματτα DK 14 (and its association with the contraries suggests *phusis*$_1$)—τοῦ τε διεροῦ καὶ τοῦ ξηροῦ καὶ τοῦ θερμοῦ καὶ τοῦ ψυχροῦ καὶ τοῦ λαμπροῦ καὶ τοῦ ζοφεροῦ, καὶ γῆς πολλῆς ἐνεούσης καὶ σπερμάτων ἀπείρον πλῆθος οὐδεν ἐοικότων ἀλλήλοις.
35. This dynamic occurs today. In discussion with geneticists working on genetic engineering at the U.S. National Institutes of Health (NIH) when creating my co-authored book on the same, I found a similar attitude among researchers: they did not care whether the details of their cellular accounts were totally accurate (in fact, most thought that every account would be regularly revised). Instead, they wanted to be able to do something that would be beneficial to some particular case brought before them. Because of this particularity, the pace of genetic engineering progress has been agonizingly slow. See Michael Boylan and Kevin Brown (2002).
36. A famous detractor to this position is Theodor Gomperz (1922). In the ancient world, Cicero, in *On Divination*, demurred that a lack in the physical account could infer a divine explanation, cf. the Empiricists in Edelstein (1967).
37. The reason for not examining the second sense of *phusis* here is that most of the writers whose texts we still have seem not to follow this path—except in the general sense of there being a possible divine explanation for why everything *is*. But such general, second-order accounts are not as useful in recognizing and explaining how first-order accounts operate.
38. See Hankinson (1998a): 17 ff., Boylan (1983): ch. 4, Boylan (1982, 1984, 1986b).
39. Not everyone is comfortable in assigning an analogous link between the activity of magic and that of science; see Gordon (1999a) and Tambiah (1985).
40. An example of alpha in our time horizon is in *The Sacred Disease: "But this disease is in my opinion no more divine than any other; it has the same nature as other diseases and the cause that gives rise to individual diseases."* V, 1–3, tr. Jones, vol. 2.
41. Kant (1781, 1787), A782/B810-A784/B812.
42. See also Hankinson (1999a) on this approach.
43. Collins (2003): 18 suggests an example for this in *pharmaka* that denotes a 'drug,' a 'poison,' or an instance of magic—sometimes from a physical figurine. This is the sort of interaction that I am advocating in this chapter.
44. Nutton (2013): 114, cf. R. Parker (1983): 208–213.
45. E. R. Dodds (1951): 146 ff.
46. Nutton (2013): 104.
47. E. J. Edelstein and L. Edelstein (1945 [rpt. 1998]), II. 1–64.
48. Rüsche (1930): 23–46.
49. PA 650b 10.
50. Blood was thought by some to be created in the heart (Boylan 1982). The rival organ to blood creation was the liver.

51. Of course, there is no comparable sense of genetics in the ancient world. Our present understanding of genetics reinforced by evolutionary theory is completely foreign. However, some rudimentary understandings can be found—especially in the writings of Hippocrates, Aristotle, and Galen. See Boylan (1984, 1986).
52. For a discussion of this, see Dodds (1951), cf. Plato who weighs in on this question with a materialist explanation, *Phaedo* 96b: "Are living creatures nurtured when heat and cold produce a kind of putrefaction, as some say? Do we think with our blood, or air, or fire, or none of these. . ." (Grube tr.).
53. I am inclined to believe that the Homeric Writers situated the mind in the nonphysical region near the heart as per *Odyssey* 17.489.
54. Boylan (1982).
55. The mixing of material and immaterial accounts occurred on some level for many years, cf. Descartes' declaration of the pineal gland as performing this function in his *The Passions of the Soul*: I.34–35.
56. Page (1972), ll. 1478 and 1480, emendation on Fitts' translation (1947).
57. From Aristotle to Galen, many of the vestiges of inscrutable vitalism and magic were replaced by attempts at material accounts; see Boylan (2007).
58. *De Caelo* 271a 33 says that god and nature do nothing in vain. This is probably not a literal claim as in the earlier texts since Aristotle's unmoved mover doesn't *do* anything, cf. Hankinson 1998a, p. 13.
59. Hippocrates, *On the Wounds to the Head*. Littré vol. XXX section 19. Translation is my own.
60. Diogenes (1925): VIII, 83
61. Ludwig Edelstein (1929): 302.
62. W. K. C. Gutherie (1962): 342.
63. Harris (1973): 4–5.
64. The main evidence of Philolaus' date is the reference to him in Plato's *Phaedo* (61d-e). The characters Simmias and Cebes say that they had heard Philolaus in Thebes at some time before the dramatic date of the dialogue (399 BC).
65. DK fragment 13; tr. Harris.
66. Rüsche (1930): 47–56.
67. Boylan (1984).
68. Boylan and Brown (2002): 133–149.
69. Harris (1973): 8, cf. Carl Johann Fredrich (1899): 67.
70. Aëtius, DK p. 214.
71. Diogenes Laertius (1925), VIII, 30, tr. Hicks.
72. It should be clear that my argument does not require the Cos vs. Sicily/Cnidos dichotomy.
73. Theophrastus, *De Sensu* 25, DK, 24, A5.
74. Kirk, Raven, and Schofield, translation: 302.
75. This has been variously interpreted, but the basic accounts can be found in Furley (1957), Booth (1960), and O'Brien (1970). I have tried this experiment myself, and it does not work this way unless the spout of the water vessel connects to the main vessel on an angle (like Basque wine bottles). If the container has an opening perpendicular to the main vessel and the vessel is placed upside down, then the air stays just where it is because the air pressure is greater than the water pressure in the case of small openings and a liter or more of air space. The variation between these two accounts says something about the nature of 'experiment' in this era.

76. Though this constant motion is an 'up and down' motion and not a general circulation.
77. Kirk, Raven, and Schofield, translation: 311.
78. Cf. Thomas S. Hall (1975).
79. Kirk, Raven, and Schofield, translation: 311.
80. This reading of Empedocles is supported by Cicero, *Tusculan* 1.9 and Tertullian, *De Anima* 5.
81. A good survey on the classic articles on reductionism in science can be found in Boyd, Gasper, and Trout (1991): 57–98.
82. This is discussed by Boylan and Brown (2002) in Chapter Three, "Mutations, Genetic Variation and Population Biology."
83. Harris (1973): 26, cf. Rüsche (1930): 112–117.
84. On the role of magic, see Dodds (1952) and Lloyd (1979).
85. On redundant explanation, see Boylan (1983), ch. 3 and Nussbaum (1978), interpretative essay.
86. *Katamenia* is explicitly mentioned in *Aphorisms* III. 28. For accounts of dual seed theories, see the Hippocratic texts *On Generation, On the Nature of the Child and Diseases IV*, Lonie, ed. cf. Littré, vii, 470–785, cf. Boylan (1984, 1986).
87. Horowitz (1976), Preus (1977), Boylan (1984, 1986). This is echoed in *Odyssey* 8. 583.
88. Hippocrates, *The Nature of Man*, lines 8–13, my translation.
89. But even these principles involved some complication. Thales said that everything was filled with gods; this would mean that his first principle, water, was also filled with gods (DK 11A 1, cf. Arist. P.A. I, de An I, 411a 8, DK 11A 3); Heraclitus' fire was also supplemented by *daimones* (DK 22A 1; Anaximander's *to apeiron* is connected to the divine, *to theion*, DK 12 A, 15, cf. Arist Iph 203b 14); Anaximenes connected air to the divine, DK 13A. Thus, the natural philosophers combine natural and magical/divine accounts—though how to understand what is meant by a divine account is a point of some controversy.
90. For a contrast of the Cnidean and Coan writers, see Bourgey and Jouanna (1975) and Boylan (1983).
91. This argument is set out by Gregory Vlastos (1953).
92. Lloyd (1979): 158, cf. Hankinson (1998a): 28–31.
93. Astronomy was a crucible of the two senses of nature. On the one hand, it was described via mathematics, while on the other hand, it portended non-materially based outcomes. See Dreyer (1906), Lange (1994), Shaw (1977), Ragep (1990), Dean-Jones (1992), and Dicks (1970).
94. Aristotle, *Historia animalium* 511 b 23–513 a 7. A. L. Peck, translation. Re: the Polybus text, see also *On the Nature of the Bones* (Littré ix, 174–176) and *On the Nature of Man* (Littré vi, 58–60).
95. Αἱ φλέβες αἱ παχεῖαι ὧδε πεφύκασιν.
96. Αἱ φλέβες ἐν τῷ ἀνθρώπῳ ὧδ' ἔχουσιν.
97. Τὰ δὲ τῶς φλεβῶν τέτταρα ζεύγη ἐστίν.
98. Obviously, our modern understanding of this word is not what the ancients intended. In their context, any cutting of the body was a surgery—whether to remove a skin tab or dermal cyst. "Bleeding" (here venesection) was an accepted practice on certain blood vessels according to the prevalent theory.
99. For a comprehensive discussion of the possible origins and uses of bilateral symmetry in ancient science, see G. E. R Lloyd (1966).

100. See the discussion by Baltzly (2004), Boylan (1983), and Wasserstein (1962). Comparisons to today on this same issue can be found in Hudson (1997) and Van Fraassen (1976).
101. See Nutton (2008).
102. For an exposition on scientific projection, see Goodman (1955) chs. 3–4. It is commonly held that this is the position of Immanuel Kant in his first *Critique*, viz., imposing a vision upon nature. This is the transcendental method; see Kant (1787).

2 The Hippocratic Period—Philosophy versus Clinical Biomedicine

The Hippocratic Writers were a diverse lot. Whether or not the historical figure of Hippocrates can be attributed to any of the texts is a contentious philological question beyond the scope of this volume.[1] There is even controversy about the traditionally recognized groups within the Hippocratic Corpus: for example, the writers from Cnidus and Cos (others have added the Italian/Sicilian writers as well).[2] Further, there is the problem of Anonymous Londinensis. Various "Hippocratic" positions are discussed by Anonymous Londinensis. Hermann Diels thought that the text was transcribed by an ignorant and unskilled student for his personal use.[3] Under this account, the work was a reader of primary sources. Wellmann applied linguistic analysis that created affinities with Soranus.[4] The possibilities spin. However, these papyri are written post-Aristotle, perhaps through some complicated authorial process in which are represented (1) the scribe (anonymous) who penned the words that exist on the manuscript page; (2) the writer (perhaps Menon, a student of Aristotle), whose text the scribe copied;[5] and (3) the authorities to whom the writer refers (these citations are occasionally quite far afield, e.g., Plato is cited for doctrines that are actually Stoic).[6] With the possibility of misrepresentation, the question arises whether Anonymous Londinensis' depiction of the Hippocratic period is accurate. I will assume that it is accurate to some text or doctrine in the Hippocratic period. Daniela Manetti suggests a compromise solution: the text is more than a compilation, but less than the work by a recognizable author.[7] If we content ourselves in our philosophical analysis to looking at a body of work that contains similar concerns that represent significant changes set out in Chapter 1, this volume, toward *phusis*$_1$ and *phusis*$_3$ (with a lesser—though still present—*phusis*$_2$) during a roughly contiguous historical period, then all of these works will be fair game for our investigation. To simplify the discourse, I will make reference to all of these groups individually by the work cited when a particular methodology dealing with blood makes such a distinction philosophically useful. Otherwise, I will refer to the Hippocratic period, Hippocratic Writers, or to 'Hippocrates' in inverted commas. In any event, I will view these biomedical texts as pre-Aristotelian (except, obviously, those references in Anonymous Londinensis that have been shown to be later).

Blood will be viewed within the contexts of (a) theoretical approaches to biomedicine; (b) basic anatomy/physiology; (c) the etiology of disease, and (d) conception theory. The goal is to contrast various approaches to biomedical issues that concern blood in some essential way.

THEORETICAL APPROACHES TO BIOMEDICINE

There are two important themes to this chapter. The first is the distinction between medical theory and clinical practice. In the contemporary era, this plays out in two different medical approaches. The first is the "theory vs. evidence-based medical conflict." Under the "theory approach," physicians set out what they think to be good treatments for disease based upon philosophical/medical principles that can drive or skew clinical trials according to one's point of view.[8] On the other hand, the "evidence-based approach" requires a case-history dialogue that involves individual case histories that explore other options based upon the particularity of the clinical presentation. A perfect contemporary example of these two approaches to medical practice is the surgical treatment for breast cancer. Some theory-approach folks prolonged a needlessly invasive surgical treatment in which not only were the breasts completely removed, but the "margins" were needlessly expanded to include the pectoral muscle mass. The result was that the patient after surgery would be largely crippled in the use of her arms. This treatment was based upon philosophical conjecture. It was only later that evidence-based medicine won the day.[9]

The second distinction between medical theory and practice is illustrated theoretically via the contemporary *token-type* distinction.[10] According to this lens of analysis, one may focus upon *type*: general descriptions about kinds of things. Or one may talk about the *tokens* Sally or Jemal—but these are individuals.[11] Individuals, as such, provide limited scientific value. For science (*episteme*), according to Aristotle, is not about individuals, but about either species (*eidos*) or genera (*genos*).[12] This is because the type does not change as tokens do. The timeless type can be *defined*. This definition is timeless. Individuals can vary by the more and the less. They cannot be defined. The type is present in the individual (token), which evinces a sort of participation with the type. However, no individual matches the type, as such. Anyone who has taken a course in taxonomy will readily understand this. One looks at plant or animal samples with a guidebook that is an approximation of the type for that species. No sample (taxon) is exactly like the type; nevertheless, we classify it within group-a (for example).

Logically, there is another distinction between type and token. In modern logic this is the modal operator of necessity ($\Box$). Necessity ($\Box$) only attaches to the type. The token possesses the modal operator of contingency ($\Diamond$). So the question then arises in contemporary philosophy of science: what is the standard of a scientific explanation? Do we aspire to timeless, necessary laws (□)? Or something else (◊)? Does the type or the token drive scientific understanding and explanation?

In post-Newton philosophy of physics, it is the former. 'F = ma' is supposed to be universally true of all times and in all circumstances. Thus, the modal operator of necessity covers the claim for any instance brought forward: any x (where x is a variable) will fall under the operator of necessity. On the flip side, if one were to find an instance of '$F \neq ma$,' then the "law" would be falsified.[13] Among the Logical Empiricists in the twentieth century, science aspired to the certainty of necessary types. This was called the deductive-nomological model (D-N). As mentioned earlier, the Aristotle of the *Posterior Analytics* would seem to have been at home with such an account (though there is some disagreement on his practice in this regard).[14]

However, beginning in the 1970s there emerged a rebellion concerning the □ quantifier regarding biology. This is because of the modern acceptance of evolutionary theory in one of its several guises. Evolutionary theory embraces variation, and some degree of chance and spontaneity with are at odds with the "lawlike" statements and the radical reductionism that was needed to maintain the D-N model.[15] This has the effect of taking the tokens more seriously. Variation within the species (a dogmatic given in modern biology) requires that the ◊ replace the □ in scientific biological theory. In biomedicine, variations among patients have ushered in an exalted role for the tokens. For example, it was discovered that individuals presented differently when having a heart attack (not all homo sapiens present in the same way).[16] This variability increases in the practice of homeopathic medicine that takes individual circumstances into account with a holistic view of the patient. All of this points to an elevation of the token (individuals viewed in their particularity) in the practice of medicine. For any explanatory scheme to be accepted as a science in modern philosophy of biological science, the tight prescriptions of Logical Empiricism and its D-N model (the 100% necessity falling under the □ operator) are discarded for something less: ◊. But how much less? In contemporary pharmaceutical testing, for example, we trot out ordinary statistical modeling that describes the anomalies (call these the 'null set'). If we can prove the null set between 0.5% and 5%, it is generally good enough for most contemporary theories of pharmacology.[17] In more qualitative language, we say that ◊ means "for the most part." Throughout his biological works, Aristotle used the expression *epi to polu* to describe this sensibility.[18]

It is doubtful that few ancient practitioners were committed to anything like the D-N model of absolute □ necessity. This is probably because even though the accounts might be materially justified (*$phusis_1$*) basing their explanation on water, air, seeds, four roots, atoms, etc., they did not have a sequential causal account. In the modern era (post-seventeenth century), such a sequential causal account is termed 'mechanical.'[19] Except for rare examples, like Aristotle's automatons, this is not the sensibility of ancient medicine.[20] Broad concepts such as hot, cold, wet, and dry offer a quasi-causal account. However, they do not specify the sequences as they causally emerge: explanation-type α: (◊) [{(x => y) · z} => q]—which can be read that if x causes y and when this event is completed and conjoined to z, then q generally results. I will take a as a standard against which to judge emerging biomedical science in the ancient world. It is temporally sequential. It demonstrates a general paradigm for a *process* by which significant events occur. Though it is *epi to polu*, ◊, it strives to create a sequential, causal, material account. There is a second explanation-type. Let us call it explanation-type β. Explanation-type β is an emergent process that embraces the ontology of *phusis*$_2$ (when the *explanans* is otherworldly or based upon magic) or *phusis*$_3$ (when the *explanans* is constrained by skepticism$_3$—[see Chapter 4, this volume, on the types of skepticism]). The modality of explanation-type β varies according to its variety (c-e, Figure 2.1).

However, this is still not sharp enough. Though the □ necessity has been mostly ruled out in explanation-type a in biomedicine (thus the ellipses), it may be a perfectionistic goal to which some authors aspire—but in practice, it is ◊ modality that rules the day. Still, we need a distinction between different sorts of contingent modalities. If we conceptualize this modal realm as a continuum, then the end points would be a ◊ modality that aspires to □ necessity (D-□, that is, technically separated from the continuum as such). Let us call this variety of ◊ modality "D-◊" (short for deductive-nomological leaning ◊ modality) and "E-◊" (short for empirically leaning [anti-hard scientific covering laws: D-□]).

It is important for the reader to note that I am arguing that Figure 2.1 represents a continuum of scientific explanation about the world—particularly in biomedicine. It reflects some of the same epistemological dynamics set out earlier between *phusis*$_1$ and $_2$ versus *phusis*$_3$. Because this is a deeply set epistemological way of viewing the world, it overlaps to other areas of philosophy as well, for example, the particularism debate within contemporary ethics.[21] It is also my contention that offspring of this continuum exist today in biomedical research and practice.[22] On the D-□ imaginary side, one dreams of utopian certainty in a realm (biomedicine). On the D-◊ pole, one emphasizes an underlying science in *principles* that are very probable, because they are supported by a general theory of science and of the body. On the E-◊ pole, one emphasizes clinical observation and the particularity of each patient's responses. It is *almost* as if each patient were sui generis (though not quite).

In the Hippocratic period, this distinction was also evident. One example of what I'm calling E-◊ is:

> Another disease: an intense head pain; vomiting of bile, difficulty urinating and delirium. The patient has pain from an excess of heat in

D-□. . . . D -◊ a---------------- b ------------------ c ---------------- d ------------------ e E-◊

Figure 2.1 Modal Continuum of Explanatory Structure within Biomedicine

> the brain. He is delirious when the blood in the brain is heated and is caused to move violently about by bile or by phlegm. He vomits bile (this liquid being moved in the body and the head), because of the heat, draws itself inward; and on the one hand vomits the thickest and on the other the thinnest is drawn in. One urinates by the same causes just has been said earlier . . . by the seventh day he often dies or if he makes it to the seventh he can expire on the ninth or the eleventh at least. (*Epidemics* II. Littré 7, p. 11, 26)

On the Figure 2.1 scale, I would rate this as between "c" and "d." There is an emphasis upon description of symptoms, but some general causal elements are mentioned: the material behavior of heat (*phusis*$_1$) that is also indirectly theory related—because the causal (responsible entity) *aitia* is set out. Blood in the brain is overheated by bile and phlegm, which blocks its somatic flow. This creates a multistage explanation-type a event. But the details are vague. This constitutes a mixture of the two poles with some definite influences of D-◊.

Another example of this ambivalence in the "b" to "c" range occurs in *On the Nature of Man* (*peri phusios anthropou* V: 1-VI: 10) in discussing the physical appearance of blood:

> I promise to show that the [the physical components] of man are eternally the same by nature [*phusis*$_1$] and by scientific law [*nomos*] . . . phlegm is unlike blood (*haima*), which is also unlike bile . . . this is because they are not warm, nor cold, nor dry, nor moist in the same way . . . there are some who think that man is composed of a single element . . . such as seeing a man who is cut and bleeding profusely [and dies] as that bleeding that caused the death indicates that blood comprises the soul of man.

The author then goes on to show exceptions to these so-called lawlike pronouncements. But what this passage shows is an acceptance that the contraries (hot, cold, wet, dry) have scientific status as such. They are used to help judge whether blood contains the soul (*psuche*) of a human. The question about soul is really about who we are. This is the most essential *philosophical* question that has ever been set forth. Who are we? Upon this answer the logical follow-up is: "How can we be more or less excellent (*arête*) according to the *ergon* (a functional condition)?" This is but one example of how questions of philosophy get into the picture.[23] Sometimes, this intrusion can lead to seemingly irrelevant nonclinical explorations. For example, in *Regimen* I.3: 1–11 the writer says:

> Now all animals, including man, are composed of two things, different in power (*dunamin*) but working together in their use, namely fire and water. Both together these are sufficient for one another and for

> everything else, but each by itself suffices neither for itself nor for anything else. Now the power each of them possess is this. Fire can move all things always, while water can nourish all things always; but in turn each masters or is mastered to the greatest maximum or the least minimum possible. (Jones, tr. [my parentheses])

This is reminiscent of the pre-Socratic philosophers who set out one or more general material principles that ruled nature. In this case, we have two principles that in their form in cooking (fire) and in drinking (water) are complementary. Water can put out fire. However, large fires are somewhat immune to being extinguished. On some meta-level, the account of *phusis*$_1$ aspires to D-□, but because there is no detailed explanation-type α, the D-□ has to be discarded in actual practice for D-◊. The advantage of this D-◊ account is that it has broad explanatory power. This represents what has been generally considered to be a positive trait in evaluating scientific theories.[24] The downside is that it runs the risk of being too abstract and divorced from the phenomenon it is trying to explain. This abstraction is too general. Abstract principles are one side of the philosophically based approach to medicine. On the Figure 2.1 scale, I would situate it either between "a" and "b" or off the chart toward D-□.

A move in the opposite direction toward the extreme E-◊ is represented by:

> The woman suffering from angina who lay sick in the house of Aristion began her complaint with indistinctions of speech. Tongue red, and grew parched.
> *First day*. Shivered, and grew hot.
> *Third day*. Rigor; acute fever; a reddish hard swelling in the neck, extending to the breast on either side; extremities cold and livid, breathing elevated; drink returned through the nostrils—she could not swallow—stools and urine ceased.
> *Fourth day*. General exacerbation.
> *Fifth day*. Death (*Epidemics III*, case 7, Littré 7: 160. Jones, tr.)

Here we are faced with a variety of symptoms that seem to be set out on a very idiosyncratic sample space. The treatment (providing liquid nourishment that is not accepted by the body) is based upon what has worked with the practitioner in the past on a kind of "trial and error" method that is decidedly situational. In this case, the treatment does not work. The patient regurgitates the fluid and loses the ability to swallow, making dehydration more acute.

Since the writer has no theory of disease or of the operation of the human body, when the only known remedy to this practitioner (providing fluids) fails, then there is nothing further that the practitioner can do. There is no back-up plan.[25]

Another example of E-◊ at the extreme (between "d" and "e") is found in *Epidemics* I from either the same author or a kindred spirit. The intent is merely to highlight various "constitutions" (*katastasis*)—times of the year or other external factors that can be identified in a loose way to be connected with the medical condition.[26] As in *Epidemics III*, there is a casebook approach that seems to be aimed at the practicing physician so that he might be able to track the stages of certain illnesses with a view to prognosis. A key common thread is identifying the crisis (*ekrithe*—where the patient has a chance at recovery). In many of the fourteen fully cited cases and the twenty-eight more briefly cited cases, the crisis plays a prominent role. It is here that the last hope of recovery is present. However, the track record for recovery is not so good. Of the total forty-two cases, twenty-five (60%) end in death. Treatment (aside from keeping the patient comfortable) is not really explored in these cases. This is because it would depend upon theories. And because they are absent, it is clear that this physician eschews the D-◊ approach. This is a bioscience of observation and the elevation of the naïve, empirical approach. Instead, the emphasis is upon *diagnosis* and *prognosis* without any conceptual link. A day-to-day account is given in this casebook. Most cases involved gastrointestinal problems or excretory complaints accompanied by fever. The physician in this case is not a healer, as such, but a palliative prognosticator. In fact, in the formal case #14, it is the patient's going to the temple of Hera that affects the cure, $phusis_2$. This harkens back to the same tensions set out in Chapter 1, this volume.

Galen, who attributed such methodology to the Cnidian Writers, comments on the E-◊ extreme approach:

> The point is that they [the Cnidians] looked at the varieties of symptoms which change for many reasons and failed to consider the specificity of the dispositions. Hippocrates, on the other hand used (for their discovery) a method which permitted one to find a number of diseases. . . . Hippocrates censures the Cnidian physicians for their ignorance of the genera and species of diseases, and he points out the divisions by which what seems to be one becomes many by being divided. (*Corpus Medicorum Graecorum* 5.9.1: 121–22 [my brackets])[27]

The point that Galen makes against the E-◊ standpoint is that the extreme Empiricist approach of this methodology is severely lacking for biomedical science. Hippocrates of Cos is brought forward as one who espouses the importance of a theoretical structure that can guide the physician in his assessment of the patient. This would suggest the contrary position of D-◊.

This sort of approach can be found in *Ancient Medicine*:

> All who, attempting to speak or write on medicine, have assumed for themselves a postulate as a basis for their discussion (*hupothesin autoi autois hupothmenoi to logo*)—heat, cold, moisture, dryness, or anything else that may fancy—who narrow down the causal principles of diseases . . . obviously blunder . . . because they blunder [about] what is an art (*techne*). (*Ancient Medicine*. I:1–4, Jones tr. [my brackets and parentheses])

The *hupothesin* is a laying down of a possibility that is open to dialectical examination. This is a philosophical tactic. Plato in his divided line sets this out at the philosophical approach at the penultimate level. [28] It is here that the authors engage, as philosophers of biomedicine, if they want to establish a *techne*—a functionally based art that is oriented toward explanatory-type α.[29] At the end of *Ancient Medicine*, the author sets out his general theory of balance so that there is established a general account of biomedicine based upon a balance of humors and their properties (such as sweet, bitter, astringent, or acid).[30] These are properties that attach to the particular humor in question, even as properties attach to variables in modern truth functional logic: Fx (where F is a property that attaches to x—which may also possess other properties).

The humor theory that is set out in *Ancient Medicine* has direct ties to philosophical problems within Plato (concerning his attack on Protagoras' methodology in the *Theaetetus*) and Aristotle (in his use of balance within the *Nicomachean Ethics*) as has been documented by Paul Demont.[31] This is in step with the inclination among philosophers for modal necessity. What is characteristic of the D-◊ accounts is first that they eschew *phusis*$_2$ accounts in favor of *phusis*$_1$ or $_3$ (unlike the *Epidemics* I passage earlier). A famous enunciation of this comes from *On the Sacred Disease*:

> This disease [epilepsy] is in my opinion not more divine [*phusis*$_2$] than any other disease. It has the same nature [*phusis*$_1$] and cause as any other disease. (*On the Sacred Disease*, 5:1–5)

As one moves toward the philosophical pole of Figure 2.1 (D-◊), one commits to more general theories that supersede the immediate clinical experience at hand. *This* patient is an instance of a generally true, quasi-natural law.[32]

At the end of the day, there will always be some tension between scientific explanation that seeks lawlike accounts on the order of D-□/D-◊ and those that present E-◊. But even among those advocating ◊, there are those who will scale up or down what "for the most part" might mean. This will be an important tension in ancient Greek biomedicine just as it is for contemporary philosophy of science.

BASIC ANATOMY/PHYSIOLOGY

In order to proceed in our philosophical inquiry, there are several key terms that begin to take on specialized meanings during the Hippocratic period. I will set out a few of these here with brief comments.

Haima, ichor, puon. The most common term for blood is *haima*. This refers to a red fluid that is in blood vessels, *phlebes* (sometimes *phlebia* for minor blood vessels). There is no distinction between veins and arteries. The word *arteria* is not used for arteries because in common usage it refers primarily to the windpipe.[33] I accept the general account that the first biomedical writer to make the distinction between arteries and veins as blood-carrying vessels is either Praxagoras (very late fourth-century Cos physician and son of a man of the same name who was himself said to be a pupil of Hippocrates) [34] or more likely by Herophilus. It should also be mentioned that *neuron* here can also be put into this general category. Although there are few macro visible nerves (Alcmaeon's discovery of the optic nerve being an exception in the pre-Hippocratic Writers),[35] and these were sometimes confused with blood vessels. Though nerves are not hollow, if one accepts the possibility of the emerging theory of pores, it might make perfect sense for nerves to be the conduits of blood as well. However, this is not the case for tendons and ligaments. Because these are situated within a very limited domain (generally between two bones or around a joint), and because they have no proximate connection to the "thick" or "hollow" blood vessels, they were generally not thought to be conduits of blood for nourishment or vitality.

Haima is generally said to be red, probably by the evidence of wounds.[36] However, some writers speculated on *why* it is red. In our modern context, we account for this because of the binding of oxygen to the heme. However, in the Hippocratic period, there was speculation that the reason for the red color was because people drank red wine.[37]

Ichor has become more complicated in its usages since Homeric "blood of the gods" by the addition of mundane usages. These include the watery part of animal meat (meat juices),[38] the watery part of the blood (serum), and the watery pus that comes from wounds and ulcers.[39] In some instances, the watery part of the blood becomes another humor.[40] But following Jouanna and Demont, the earlier Homeric usages do hang on, as in the case of Clytemnestra's speech in the "Agamemnon" v. 1480. But it doesn't stop with Aeschylus—Jouanna and Demont cite lingering references (by extension) all the way to John Milton in *Paradise Lost*.[41] This reading of one of the senses of *ichor* seems correct to me. There are two usages of *ichor*: (1) the Homeric sense of divinity, *phusis*$_2$, and (2) the more mundane, purely material usages of *phusis*$_1$. This maintains two of the three senses of nature set out in Chapter 1, this volume.

Puon is pus. It is the excretion from a wound or ulcer. As mentioned earlier, it is sometimes conflated with *ichor*. The key thing to note here is that blood itself is watery. However, when there is a wound or ulcer, a clear fluid separates itself. In modern medicine, we call this the plasma. In ancient medicine, it was either *ichor* or *puon*. This means that blood has a more complicated structure than a mere uniform part (see Chapter 3, this volume). It is important to note that seeing blood as a complicated structure is extremely important in evaluating the biomedicine of these writers. As noted in Chapter 1 of this volume, air can get mixed in the blood with various results. The nature (*phusis*$_1$ or $_2$) of the blood is of utmost interest in our exploration.

Aer, aither, phusa, pneuma. These are all terms that refer to some version of 'air.' They are not consistently used in any technical sense. Since air is breathed into the nose (and possibly through pores in the skin), it has a very real role in the vital account. If someone were to block one's ability to breathe, he would die. When people die of other causes, they cease to breathe. Some of the Hippocratic Writers (previously identified with the "Coan Writers"—especially the author of *On the Sacred Disease*) picked up on this argument and also cited the brain as the seat of reason and the soul (though not always of emotion).[42] Other biomedical authors of this general period, such as the writers of *On Places in Man, On Disease*, and the fragment *On the Heart*, were keener to emphasize the role of blood over air (and its proximity to the brain via the nasal entry canal) and therefore made the heart the seat of the soul and reason.[43]

However, there are some differences in the way these four words are used. *Aer* is a general term for the atmosphere we encounter in daily life—altered sometimes by being a little dry or moist.[44] Some ancient writers separated *aer* and *aither* as a pair, with *aer* being below and *aither* being above. This dichotomy, that was set out by some ancient historians, is criticized by Liddel-Scott-Jones (the principal classical Greek dictionary) as being inaccurate. *Aer* is often associated with one of the four roots of Empedocles: air, earth, fire, and water. But Empedocles often used *aither* interchangeably with *aer*: DK 71, 98, 109, 115. Aristotle thought that Anaxagoras believed *aither* should be associated with fire: DK 2 on the formation of the universe. And Aristotle also posited *aither* as the fifth element that constituted the super-lunar realm that was beyond generation and corruption and possessed the property of eternal, circular motion: *peri ouranou* (*On the Heavens* 269b 31, 270b 23).

In *On the Sacred Disease* the writer discusses air flow within blood vessels as a physical account (*phusis*$_1$, explanation-type α). Under this etiology, when the air (*pneuma*) is blocked in a minor blood vessel, normal physiology ceases. An example of this is when we are seated or lying down and we press against a minor blood vessel, we lose feeling and function is impaired (VII). When we apply this principle to the brain, there are serious effects:

> The air (*pneuma*) that goes into the lungs and the veins (*phlebes*) is of use when it enters the cavities and the brain, thus causing intelligence and movement of the limbs, so that when the veins are cut off from air (*aer*) by the phlegm and admit none of it [the *pneuma/aer*], the patient is rendered speechless and senseless (X: 20–25, Jones tr. [my brackets and parentheses])

Thus, when the explanation-type α plays out and blood vessels in the brain are blocked by phlegm (another bodily fluid of high viscosity), it is just like the case when one sits or lies in such a way with one leg over the other that numbness occurs in a limb. However, because the site of the blockage is the brain, major consequences occur: epilepsy. This account is to be viewed in contrast to other contemporary practitioners who invoke *phusis*$_2$ accounts:

> My opinion is that those who first attributed a sacred character to this malady [epilepsy] were like the magicians, purifiers, charlatans and quacks of our own day. (II: 1–4, Jones tr. [my brackets])

As we have just seen earlier, *pneuma* can refer interchangeably to *aer*. However, as a vital, material component, it must obtain its power somehow. The *phusis*$_1$ explanation-type α candidate for this is heat. Heat is known by any cook to animate a soup stock. It can change the character of food in cooking. Also, heat is another critical difference between a live person and a dead person. It is thus logical to view heated air as possessing special properties necessary to life and the physiology of the individual.

The process of taking in external air and transforming it into internal body, energized air (sometimes referred to as *pneuma*) involves the physical intake of air through breathing or breaths.[45] *Phusa* is sometimes limited (as per the author of *peri phuson, On Breaths*) to the air within the body. In Chapter III the author says that man is nurtured by three essential substances: food, drink, and wind (*pneuma*). Outside of the body, wind is called air (*aer*) [here contextually synonymous with *pneuma*] and inside the body it becomes *phusa* and acquires a power (*dunamis*). This air mixes with the blood to create a physiology that can both provide nourishment to the body and fulfill the functions of a nascent neurological system—all under the control of energized air within the blood.[46] In Chapters VI and VII the author extends the reach to *common* and to *particular* diseases. The common diseases are like the plague. They are spread through groups of people breathing bad air.[47] In this text, the bad air commonly breathed is termed *pneuma*.

In the case of particular fevers (*puretoi*), these generally come from diet (which is more individually variable). Lloyd suggests that even in eating *pneuma* gets mixed in—as per the well-known case of belching as a result of eating or drinking (cf. the odd, disconnected Chapter IX of *On the Nature of Man*).[48]

Kenos. This is a term that is often used by the atomists, *to kenon*, to explain movement and how things come to be. Anonymus Londinensis uses it within the phrases *kenos athros topos* (κενὸς ἀθροῦς τόπος), or *ho kenos athrous* (ὁ κενὸς ἀθροῦς), XXVI, 48c, XXVII, 6, 7, 38. These connect to nature's abhorrence of a vacuum. Such a vacuum could be the lack of a physical space (or place, *topos*), or it can refer to physical events having causes (akin to Leibniz's Principle of Sufficient Reason—see Chapter 1, this volume). When applied to the biomedical realm, the term refers to a vacuum. Erasistratus is connected to a methodological view that Nature abhors a vacuum.[49] This is a persistent concept in the history of science.[50] It is very similar to "Nature does nothing in vain." In each instance we have a *purposive* sense of Nature (*phusis*$_2$) that is active in the world. When there is an explanatory vacuum, it must be filled. The reason is that a comprehensive, creative god would not leave the details unfulfilled.

A purely material account (*phusis*$_1$) would be that there is some sort of tangible principle according to which a material vacuum is impossible, D-□. If one were to suck out the air from a vessel and then submerge it into water, the water would rush in (consistent with the famous clepsydra experiment set out in Chapter 1 of this volume).[51] Thought of in this way, the vacuum is a powerful principle of *dunamis* (here understood as physical exercise of power).[52] Most likely, *kenos* need not require an actual vacuum (though conceptually this is necessary for the historical concept). Instead, a blood vessel temporarily vacant of blood after its stock has regionally been absorbed into the body might be *kenos*.[53] Its emptiness and its ability to attract blood might be enough to fill the empty spaces. The process of attraction has been the subject of several studies, but these are not set up on the explanation-type α.[54] As such, these accounts are rather inscrutable; they fall under *phusis*$_2$ and$_3$.

ETIOLOGY OF DISEASE

So how is it that we become sick? What are the causes of incapacitation and death? And what role does blood play? This is a primary concern of the medical writings of Anonymus Londinensis (IV, 26-XXI, 9). Some of these concepts have already been examined. For our purposes here, let us assume that there are two overarching understandings about the etiology of disease: 1—general deductive nomological accounts: (a) exterior according to overarching philosophical principles of the structure of nature (*phusis*$_1$)—D-□ to D-◊ and (b) general interior versions of the humor theory D-◊; versus 2—specific applications (what I call clinical biomedicine that focuses upon some patient sitting before me *now*) of the accounts according to some *epi to polu* rule—leaning heavily toward E◊. This latter mode is a highly idiosyncratic personal evaluation that employs seeming intuition over and against the application of some theoretical construct D-◊.[55] This tension between the enunciation of the general—leaning toward modally necessary theory, diagnosis, and treatment—is what I am calling the philosophical approach. It will develop into what is later called the Dogmatist position.

The other approach is highly particularized and depends upon the experience of the physician. This will later develop into the Empiric position. More on these distinctions and on the development of Methodism will be presented in Chapter 4 of this volume.

The principal voice behind Anonymus Londinensis begins by declaring that disease is against Nature, *para phusin* (IV: 2). The only way that this is comprehensible is that it contradicts *phusis*$_2$ because the very notion of *against* implies a regularity that is normatively correct. Things are supposed to act in a particular way, and when they do not, then an aberration has occurred. Whenever the concept of "supposed to" is brought forward (that things should act in a particular way *kata phusin*—according to nature), then a normative element has been introduced. This means that the proper state of affairs, *kata phusin*, is right and *para phusin* is wrong. Such judgments cannot be made unless there is a general normative backdrop that *this is the way things ought to be*. The "ought" requires a foundation. In the case of nature, this could only be the creator of nature (god) or the existential mode of a perfect whole, regardless of whether there was a creation, $phusis_2$.[56] From this perspective, to be in health is good and to be diseased is bad—not just from the individual's prudential standpoint, but from a rather larger—philosophical—standpoint.[57]

From the $phusis_1$ and $_3$ standpoints, normativity in nature is an absurdity in expression. Because these perspectives are explanatory-type α oriented and there are no other actors save material and its properties, to say "should" is ridiculous. For these proponents only explanatory-type β with its emergent consequence, there is no "should": only "is" will do. Thus, from this perspective, disease and health are just two outcomes that can be understood by the explanation type-α approach. There is no *para* or *kata* nature. There are only the factual outcomes of nature. In Anonymus Londinensis this conflict between philosophy and clinical biomedicine is set out by the author:

> But as to the subject before us there is much difference of opinion. For while some have said that diseases arise because of the residues (*perissoma*) from the nutriment (*trophe*), others hold that they are due to the elements (*stoicheia*). (IV: 26–28, Jones tr.)[58]

This quotation supports the major tension set out in this chapter between the philosophers who seek general principles such as the Empedoclean element theory (often conflated with the four-humor theory) or how external air affects general health. These range in their intended authority (modality) between D-◊ and D-□. Plato's (though not a "Hippocratic Writer") short discussion of medical theory in the *Timaeus* gives another view of the philosophical pole in this debate.[59] Plato uses the cosmic truth of the four elements to mirror four analogous principles in the human body. The underlying sensibility of this sort of argument is that there are major principles that govern everything: earth, air, fire, and water. These can explain everything (perhaps with the addition of the hot, cold, wet, and dry)—including ancient medicine and accounts of the blood. We have seen earlier where this sort of orientation easily includes water (blood is a fluid) and fire (blood is warm). Air is a little trickier because it has several divisions between external, internal, energized, and inert. Earth is a placeholder for tangible substance—hardness and the degree of impenetrability. In this way, a general material view of $phusis_1$ takes over. Such a conquest seems to minimize the role of the experienced clinician. I characterize this as philosophy triumphing over clinical medicine.[60]

Another example comes from the writer of *Fleshes* who sets out that the condition of a person is necessarily connected to large D-◊ forces that also rule the cosmos.[61] In *Ancient Medicine* the writer sets out a theory balancing various internal fluids (humors, *chumon*) that have power through themselves. Each humor can become sweet (a *para phusin* condition). Next is acid at the other end of the spectrum. What is best is something in the middle.[62] The fluids are in constant flux, but they are controlled by a law-like structure that is based upon the principle of reciprocity. The author of *Breaths* states this in terms of hunger and satiation.[63] If we are hungry, then we should eat something, and this opposite brings about the middle, natural condition of satiation. Though the laws are necessarily true (D-□), the interpretation needed by the practitioner renders the modality (D-◊).

On the other side is the single medical practitioner who uses his own personal clinical experience with patients (perhaps with a little theory too) to make an idiosyncratic diagnosis and treatment recommendation, E-◊. One reason offered for this account centers around the pesky digestive system (still a diagnostic difficulty even in our modern era).[64] Instead of (in principle) regular, predictable outcomes, one is faced with nuance. Beate Gundert suggests that vague losses of *dunamis* within the digestive system because of a lack of heat to digest can create an excess of food that does not become nourishment (*trophe*) and instead becomes residue (*perittoma*).[65]

Anonymus Londinensis, once again, enters this debate: "Thrasmachus of Sardis declares blood as the cause of disease."[66] In this case it is more properly characterized as the change (*metabole*) of the blood into another sort of humor (fluid). For example, blood can become phlegm, bile, or pus. When this occurs, the normal amount of each humor is altered and there is an imbalance; q.e.d. disease occurs.

Then the principal voice in Anonymus Londinensis develops this thought: "Dexippus, the Coan, thinks that disease is caused from residues of the nourishment . . . from the powers of bile and phlegm that come-to-be from a part [only] or from the whole body."[67] What these E-◊ accounts have in common is a vague background condition that might account for the symptoms that present to the physician who, in turn, must make a very personal assessment of them for the purpose of treatment.

In *Diseases I* the author begins with an exhortation about the clinical setting: the patient and the physician. It is set in the very Heraclitan context of constant changing. The disease might begin one way and change to present in another way. Diagnosis and treatment under such circumstances might owe as much to luck as to theory.[68] Physicians are said to be guided by conjecture (trial and error). This realistic sense of the role of a clinician mirrors many internists practicing today on atypical cases. The author of this text has captured a real element of the complexity of the doctor–patient interaction that continues to endure 2500 years later.

This is not to say that there isn't *any* theory; it is just set out in an E-◊ manner. For example, in *Diseases I*. 14 the author talks about discharge from the lungs. This is said to occur due to a rupture of a blood vessel in the lungs:

> Some of the blood (*haima*) the patient expectorates immediately, but, unless the vessel closes, other blood is poured into the lung and putrefies (*sepo*) there; when this has putrefied, the patient expectorates pus (*puon*): with the passage of time, sometimes pus proper, sometimes pus charged with blood, and sometimes blood. (*Diseases I*. 14: 162, Potter, tr.)

The clinical event is described with many side possibilities. This fits with the E-◊ clinical approach. The putrefaction (sepsis) scenario is similar to the residue (*perittoma*) scenario: there is an idiosyncratic inefficiency of the biological process. In the previous case, it is because a rupture of a blood vessel in the lung. *Too much* blood makes the constituent parts break apart. The pus (serum or plasma) separates. This is verified by the patient's spitting up. The separation is caused by the putrefication. In the case of residue (set out earlier), there is an analogous process in which a toxic process occurs. Among several treatments suggested, one is bleeding the patient so that he no longer has a surfeit of blood (see Chapter 5, this volume). However, since the bleeding occurs in the arm, it is not as conceptually clear how this helps a problem originating in the lungs. The only retort is that the author of the treatise tried this treatment on occasions and it worked. Trial and error: the backbone of E-◊.

CONCEPTION THEORY

One way to conceptually unite some of the points that have been raised in this chapter is to briefly examine the role of blood in conception theory. What makes conception theory so intriguing to philosophy of science in most of human history is that so much must be surmised from the beginning and ending stages of the process: the sexual act that gets everything going and the emergent child. All the bits in between require a large amount of speculation. One way to approach this question is to simply assert that it happens, *phusis*$_3$, explanation-type β. But what about explanation-type α? This approach would combine aspects of *phusis*$_1$ and *phusis*$_2$, as Paul Demont does in his examination of *trepho* (τρέφῶ)[69] that seeks to blend observations on how the word works in an ordinary way as a device to curdle milk to another sense of *how* the reproductive fluids come together and then to grow.[70] This seems to be an attempt to move the process from the D-□ to D-◊ range of modality to one involving E-◊ dynamics. Obviously, one could write an entire book on the subject of conception theory among the Hippocratic Writers.[71] However, in our rather rapid overview, a few things can be said.

To quickly set the stage, we begin with one version of conception that emphasizes the male role. This was a very important theory in the ancient world. In Aeschylus' *Eumenides*, Apollo says, "The mother is not the parent of that which is called her child but only nurtures the new planted seed that grows. The parent is he who plants the seed" (658–660).[72] In Euripides' *Orestes*, the mother is likened to a field that is sown (552, cf. Plato's *Timaeus* 91d 1–2).[73] Anaxagoras also seems to have supported this position according to Aristotle, *G.A.* 763b 31–33.[74] I will term such position the *male single seed theory of conception*. The characteristics of this approach are that the male provides the essential material for creating a child and the female provides only a negligible portion (sort of like a terrarium or growing chamber). Under this account there would be two separate processes at work: (1) *in the male*, some potentialities for all the parts necessary to produce the offspring would be sent to the testes via the blood or some other fluid; there the potentialities would either exist side by side to express themselves sequentially after being implanted in the female (epigenesis), or they would form together in the male seed as a proto-little person (pre-formation) and then, when implanted in the female, they would simply get bigger; and (2) *in the female*, there would be a process of making the uterus as welcoming as possible to the male seed by the nourishing power of the female menses, *katamenia*. In each case, blood is crucial to the creation of new life. However, there is no material account of *why* male blood contains more *dunamis*. It is this author's opinion that here we have an example of a social prejudice slipping in by the backdoor to affect biomedical theory. Women in the ancient Greek world were socially and politically second-class citizens.[75] Thus, since men played the powerful role in society, it might make sense to some that their blood-concocted seed would be more powerful as well. Of course, there are other examples of female power in the ancient world, but this was more often moral excellence as opposed to physical fortitude.[76] More will be made of the male single seed theory of conception in Chapter 3 of this volume.

In contrast to the male single seed theory is the dual seed theory. This is the theory of choice for particular writers in what had traditionally been thought of as the Hippocratic Writers. One obvious advantage of the dual seed theory over the male single seed theory is being able to account for mixed inheritance. The male single seed theory predicts that all children are like their father, according to nature (*kata phusin*$_1$). If they vary, then it is an instance of being contrary to nature (*para phusin*$_1$). However, what material cause would create a child who resembled the mother? It is not enough to say that is simply *para phusin*$_1$. This is because if a child did not resemble the father, then why should he or she not resemble some random set of human features or some gross malformation (what we moderns call mutations)? Why should the offspring resemble the mother or some maternal relative? The male single seed theory has no credible answer to this query (see Chapter 3, this volume).

In the dual seed theory both the mother and father each produce a seed. The female seed is related to blood as it is connected to the *katamenia*.[77] The female and male seed may be the result of pangenesis in which each part of the body offers a *dunamis* into a general fluid. The general fluid may travel in the blood as it descends to the reproductive organs. Such a pangenesis account would feature the role of blood as part of the transmission agent. This is the sort of account that is most often represented among the Hippocratic Writers. One key part of the pangenesis argument is that there is a drawing or pulling by the gonads to the entire body during the sex act. The male drawing is quicker and more intense, while the female is slower, longer, and less intense.

> I assert that seed [in men] is secreted from the whole body, from the solid parts and the soft parts, and from all its moisture. And there are four kinds of moisture: blood, bile, water, and phlegm, for this is the number of kinds of substances a person naturally contains in himself . . . and women, too, ejaculate from their body, sometimes into their uterus . . . and a woman feels pleasure, once intercourse begins, for the whole time. (*Peri gones*, ch. 3–4, Potter tr. [my bracket])

The other possibility (epigenesis) would counter pangenesis. In the epigenetic account, some general concoction (*pepsis*) occurs from the blood in the reproductive organs. In this case, a seed is produced that has the power to fashion the new fetus part by part. In male single seed theories, the metaphor is of a sculptor working with difficult materials to create a statue (the embryo). Dual seed theories, among the Hippocratic Writers, are generally of the pangenesis variety. Aristotle will promote the epigenesist account in his version of the single male seed theory. In dual seed theories, there may be war imagery between the two seeds that is meant to depict the process of genetic inheritance.

The sex of the offspring might be determined by the winds (especially in Eurymachus, Parmenides, Empedocles, and Asclepiades) or by the part of the uterus in which the seed grows (Anaxagoras and Parmenides) or by the origin of the male sperm (Cleophanes). Leucippis claimed it was a pure lottery and no regulative factors could be ascribed—even as parameters.[78]

In the Hippocratic work *Peri gones* (*On Generation*), the writer offers proof of dual seed theory by the fact that both men and women experience pleasure in intercourse.[79] This pleasure is indicative of a *pulling* from the various body parts by the gonads some liquid material that contains the *dunamis* necessary for reproducing each body part. Also contained within the liquid are the four distinct fluid varieties: blood, phlegm, water, and bile.[80] In general, the female seed is moist and cold, while the male is dry and hot.[81]

The man's semen comes from the fluid that is in the body, beginning with the head and leading to the spinal cord and down to the kidneys through a passage (*hodos*) to the gonads.[82] Though the "passage" here is rather generic, it might not be too much of a stretch to compare this to a similar journey taken by blood from the writer of *On the Nature of Man*, 11.[83] Here, the writer sets out in considerably more detail the blood vessels running down the neck on each side of the spine until they reach the hips. Others go to the major organs—such as the kidneys and then to the gonads. This route is popular among several writers of the period. If it corresponds to the path of blood, then perhaps they (seeds, *dunameis*, from each part) were thought to travel together with the blood. The lack of specificity of the route of the pangenesis material renders Hippocratic conception theory into the realm of *phusis*$_3$. The process is decidedly material, but there is nothing approaching an explanation-type α.

The role of blood in adult women is even more highlighted. This is because of the menses (*katamenia*). The writer of *On the Nature of the Child* sets out this relationship in both pregnant and nonpregnant women:

> When a woman is pregnant, she is not troubled by the failure of her menses to pass, since her blood is not stirred up (*tarassetai*) as it separates off in a mass each month; rather blood passes into her uterus gently a little at a time without trouble each day, and what is growing in the uterus increases. Blood flows each day and not one time each month because the seed present in the uterus draws continually as much of it from the body as it has strength to. At that place breath (*pnoe*) passes. At first there is a small amount of breath, and little blood flows to the mother, but when the amount of blood becomes greater, it draws more blood and the flow down to the uterus increases. (493–494, Potter tr.)

The only way I can see to understand this passage along the lines of an explanation-type α account is to conclude that the flow of blood to a woman's uterus is due to a drawing by the uterus. For the nonpregnant woman, this "drawing" or "stirring up" is pulled down all at once, regardless of whether there is intercourse. This occurs once a month. When there is an embryo in the uterus, then the process changes to a regular trickle that nourishes the developing embryo. Among nonpregnant women, it is a boom or bust cycle. If the woman is after the menses, then gradually she becomes dryer and more akin to the male (whose nature is dry and hot). Under these conditions, a woman becomes more fertile.

> Women who are not pregnant are troubled when their menses fail to pass . . . if a woman is quite empty of blood, she becomes pregnant whereas if she is full she does not. For when the uterus and vessels are empty of blood, women conceive children. (495, Potter tr.)

If one were to set a day length in the twenty-eight-day cycle for "drying out" or becoming empty of blood as a couple of weeks, then the writer of *On the Nature of the Child* has the fertility cycle generally correct. What he misses is the end of the cycle (before menstruation), when the female is again non-fertile. Because the language is set out in lawlike terms, these writers seem definitely D-□ oriented. This is supported by giving an account based upon balancing moisture (the female nature) with dryness (the male nature). When the uterus is properly dried, then fertility can result. This formula aspires to the explanation-type α. Of course, this is what we would expect in a D-□ or D-◊ account. This is because in order to propose a lawlike account, one needs to give some specification for the proposed biomedical laws. This requires a process with some structure that would allow the author to claim some regularity in its execution. This is what the deductive-nomological orientation is all about. The Empirics, in contrast, struggle to maintain the clinical individuality of each person as (at the extreme) *sui generis*. This is a perennial tension in biomedical theory and practice that carries on to our present time.[84] How this struggle continues will be developed in the subsequent chapters.

NOTES

1. Edelstein was skeptical that any of these texts were written by the historical Hippocrates: Ludwig Edelstein (1931, 1939). Some of these skeptical concerns are also raised by Philip J. Van Der Eijk (1999, 2005a). For a contrary point of view, see Pohlenz (1938): 3ff. For a methodological overview of these attribution practices, see Asper (2013). For the most part, I will follow G. E. R. Lloyd's (1975) account. Lloyd's view is one of general cautious skepticism. One can find various similarities of style in various works, but it is not enough to claim common authorship beyond just a few paired texts. The reader is especially enjoined to look at Lloyd's thorough bibliography on this question.
2. Some of these arguments are laid out using Anonymus Londinensis (137) by Danielle Gourevitch (1989). Nothing is conclusive vis-à-vis whether there was a "school of medicine" in a strict sense, but the author thinks the notion has merit.
3. Hermann Diels, ed. (1952): 412–413.
4. Wellmann (1922).
5. Jones (1947): viii; 4. See also Mejer (2002).
6. *Anonymus Londinensis*, XIV:16–32: 252–253.
7. Manetti (2013).
8. Sometimes this strong belief in the truth of a theory (be it philosophical or medical) can lead to research fraud, as in the case of the late Dr. Roger Poisson formerly of St. Luc's Hospital in Montreal. He was so sure of the truth of lumpectomy as a surgical treatment for breast cancer that he falsified data in his clinical trials (Fisher and Redmond 1994).
9. A good example of this is cited by Otis Webb, MD, vice-president of the American Cancer Society in his 2012 book on the subject—especially Chapters 18 and 22. Webb notes that the prevailing medical theory at the time set out a surgical treatment for breast cancer that was based upon the prevailing theories. This procedure required the removal of not only the breast, but the entire pectoral muscle mass. This meant that survivors could no longer use the arm on that side of the body with any proficiency. The theory held fast within the medical community because it had always been done. In this case, the "Empirics" tried something different and used their clinical results to change the range of breast removal to the breast itself or later even the lump. The force of evidence-based medicine Dr. Webb contends is a positive driving corrective in medicine.

10. The type-token distinction is variously represented. My use of the distinction owes its source to Paul Grice (1969) and Willard Van Quine (1987). For a straightforward account intended for students, see David Papineau (2013): 19.
11. One account that supports the tension between the individual and the type can be found in Véronique Boudon-Millot's study of στοχάζεσθαι, which is a common term in the works of Galen, but also occurs six times in the Hippocratic Corpus. Boudon-Millot makes a connection to the writings of Plato and Aristotle with respect to their emerging connection to 'conjecture' or 'guessing' that moves between the individual and the type when the practitioner is in a clinical situation in which ignorance rules the day. In this case, 'conjecture' or 'guessing' may be the best way forward. See Véronique Boudon-Millot (2005).
12. I have always argued that *eidos* when used by Aristotle to refer to a kind is in the context of a larger class, *genos*, Boylan (1983): 50–59. A similar argument is made by Pierre Pellegrin (1982): ch. 2. The view that the terms are interchangeable is made by David Balme (1972): 120–121.
13. A famous example of this is Karl Popper (1959[1934]): ch. 4. For some famous criticisms of this position, see Lakatos (1970): 170–196.
14. I have made the argument that the method set out in *Apo* is not carried out in the cannon of Aristotelian biological practice—such as *P.A.*, *H.A.*, and *G.A.* in Boylan (1983): 18–21, cf. Jonathan Barnes (1982): 37. For the other side, see David Balme (1961): 195–212.
15. A sample of the writers I'm referring to here include David Hull (1974): ch. 3; Alexander Rosenberg (1985): chs. 3 and 5; George Williams (1989): 182–185; Francisco Ayala (1989): 187–197; and Kim Sterelny and Philip Kitcher (1988): 339–361.
16. Gender differences in the presentation and treatment of heart disease are now generally recognized. See National Library of Medicine (2002): 126; Shanna S. Roberts (2008): 48–49; D.K. Quek (1990): 185–188; and Cristina Vassale et al. (2008): 608–613.
17. For an overview of the inductive logic involved here, see Ian Hacking (2001): ch. 18.
18. The proximate declarations on this doctrine occur at *Phys.* 198b 35–36; *Apo.* 87b 19–22; *Metaph.* 1027a 20–21, cf. Michael Ferejohn (1991): 119–123. This concept was far reaching for Aristotle that extended to the *Ethics* as well—see Michael Winter (1997): 163–189.
19. Clocks and designs have been a key metaphor for understanding a mechanical structure of the scientific realm. This was especially evident in the seventeenth century. Rene Descartes compares the human body to a clock after a short discussion on machines and automata (mechanical puppets): René Descartes (1911 [1637]) *Discourse on Method*: 116–117. Robert Boyle used the same metaphor after observing the impressive large clock on the Strasbourg Cathedral, Robert Boyle (1980 [1660–1691]): 33, 71, 129, 160–162, 162, 169–170, 185–186. One famous example that combines the mechanical watch with teleological design is William Paley's argument for theistic creation: William Paley (1867), *Natural Theology: Or Evidences of the Existence and Attributes of the Deity Collected from the Appearances of Nature*: 1–13, cf. the Cambridge Platonists (e.g., Henry More (1671) *Enchiridion Metaphysicum; sive, de rebus incorporeia succincta & luculenta dissertation I*: 35–70, who are not as explicit in their metaphorical usage. For a defense of some usages of *mechanical* in the ancient Greek world, see Berryman (2009): chs. 1 and 2.
20. "And it is possible that A should move B, and B move C, and that the process should be like that of the 'miraculous' automatic puppets: the parts of these automatons" (G.A. 734 b 10–18; cf. 741b 9; M.A. 701b). Though Berryman (2009): ch. 5 makes the case that it was more common than most

commentators have thought. This is especially important concerning *pneuma* and "pneumatics."

21. The particularists represent the E-◊ branch as opposed to the D-◊. In the ancient world, ethical particularism was characterized by the genre of *aphorisms*; see Hippocrates, *Aphorismoi*. These short, pithy sentences aspire to set out a narrow, particular truth. This is quite different from Aristotle's *Nicomachean Ethics* V., for example, that seeks to set out universal criteria on justice, cf. Plato's *Republic* that has a similar aspiration. Thus, regarding ethics, the aphorisms would connect to E-◊, while Aristotle and Plato would connect to D-◊. This same dichotomy in ethics exists today. Jonathan Dancy (2004) aspires to E-◊, while naturalist philosophers such as Kant aspire to either D-◊ or D-□, depending on the interpreter.
22. For a discussion of this dynamic, see Boylan and Brown 2002—which also sets out some objectors from the "evidence-based" practitioners (who are often reticent for experimental treatments): see part 2.
23. Of course, there are other ways that philosophy and medicine in the ancient world were said to interact. My approach is to use a history of ideas regarding philosophy as I take various interpretative lenses and apply them to ancient medicine. Vivian Nutton (2013): 45–50; 116–122 uses philosophers of the period and compares them to medical writers of the period, first with the pre-Socratics and then with Plato and Aristotle, cf. Paul Demont (1998) and Jacques Jouanna (1980). Philip van der Eijk (2008) takes a similar tack. Van Der Eijk's position is that there is not a hard line between the ancient philosophers and those who were interested in or even practiced medicine; cf. Michael Frede (1986). This is certainly true, but it is a different sort of exploration of philosophy and medicine than I am engaging in. Finally, a third sort of approach is evidenced in Geoffrey Lloyd (1963) who takes a particular text and a set of complaints by the ancient author and imaginatively tries to reconstruct and examine the significance of these. R. J. Hankinson extends this to reflections on the history of philosophy. The combination of Lloyd and Hankinson is most similar to my own approach.
24. It is certainly hard to say whether these sorts of pronouncements, as per *Regimen* I, are D-◊ or D-□, but it will be the position of this book to remove D-□ from the consideration table in biomedical explanation. It still could be considered in astrology or any other mathematically based science.
25. In fact, introducing fluids to someone with a fever is ceteris paribus a good thing if the cause of the fever is one that the body can fight on its own. Since we require regular hydration, the default treatment is just fine. However, because this practitioner is so biased toward E-◊, there is no second go-round (because there is nothing to base it upon).
26. This is also the general concern of *Airs, Waters, and Places*, however; in each, there is the obvious danger of *post hoc ergo propter hoc*. A conjunction of physical events does not constitute causation—the responsible party, *aitia*.
27. Müller (1981): 776.
28. Plato classifies the laying down of hypotheses as the penultimate level of knowledge acquisition in his "divided line" *Republic* 509d-511e.
29. Plato was relentless in demanding explanatory-type α accounts. For example, in the *Protagoras* he has Socrates doggedly ask for the *techne* for excellence (*arête*) that Protagoras, the sophist, was trying to peddle. This is repeated over and over in other dialogues as Socrates cites examples of productive crafts, like shoe making, that are clearly sequentially based material endeavors.
30. *Ancient Medicine*, 24.
31. Paul Demont (2005).
32. I am following here what I take to be the conventional view espoused by Nutton (2013): 65, that material explanations are to be preferred to *phusis*$_2$

explanations—though conventional religion is not being attacked, as such. However, others sense a strong *phusis*$_2$ flavor in the way the denunciation occurs—see Jeanne Ducatillon (1977): 180–185; and, to a lesser extent, Phillip J. van der Eijk (2005a): 45–73.

33. Harris (1973), 24.
34. Some have traced this as far back as Euryphon of Cnidos: Max Wellmann (1901): 91; cf. Carl Johann Frederich (1899): 68.
35. Chalcidius makes this claim in his commentary on Plato's *Timaeus*, D-K, vol. 1, 212: 20–37.
36. Of course, there are the usages of blood as black as in *Iliad*, XX, 470, XXI, 119. But perhaps these usages mix the dark drama of death with the presentation of blood.
37. Jouanna and Demont (1981): 203.
38. See Hippocratic text, περὶ καρδίης, 11, Plato, *Timeaus* 83c, Aristotle, 52b 2.
39. See Jouanna and Demont (1981): 201 who cite *Diseases of Women*, c.64, Littré VIII 130, 22 in relation to ἕλκος (in the sense of ulcer), cf. περὶ ἑλκῶν, ch. 27, Littré VI, 430, 22. It should be noted that Jouanna and Demont believe that the primary usage of "blood of the gods" continues through the Tragedians and lingers beyond—even to Milton.
40. Littré, V. 697, 699, and VI 63.
41. Jouanna and Demont (1981): 208—"Then Satan first knew pain . . . from the gash/ A stream of nectarous humour issuing flowed/ sanguine, such as celestial spirits may bleed" *Paradise Lost* VI: 327–334.
42. Anonymous Parisian of Fuchs's *Anecdota medica* declared that Hippocrates proclaimed the brain to be the Acropolis of the body, R. Fuchs, "Anecdota Medica Graeca," *Rheinisches Museum* 49 (1894): fragment 1. Aristotle quotes Polybus, the pupil and son-in-law of Hippocrates, as making the brain primary because it is the origin of the most important blood vessels in the body (a claim that Aristotle does not agree with)—H.A. III.3, 512b 12–513a 7, cf. *de natura ossium*, ch. 9 (Littré IX, 174–176) and *de natura hominis*, ch. 11 (Littré VI, 58–60).
43. That these writers were from the Cnidian school has been disputed: I. M. Lonie (1965). According to Lonie's argument, the only clearly Cnidian texts are Περὶ παθῶν, Περὶ νούσων, and Περὶ τῶν ἐντὸς παθῶν. On the other hand, *On Regimen in Acute Diseases* has the character of E-◊ that this author associates with writers attributed to Cnidos.
44. An example of emphasizing environmental air is in *Airs, Waters, and Places*. Here I am following Lloyd (2007).
45. This can also be the cause of disease, with or without religion (*phusis*$_2$ and *phusis*$_1$). For a discussion of this that examines an understanding of causes consider the Athenian Plague during the Peloponnesian war that employs a primary sense of causation (without religion). Demont (2013) points out that crowded living conditions during the Peloponnesian war (when Pericles brought the rural population into the city in crowded conditions that fouled the air) provided a material account of an epidemic. However, the secondary cause of *phusis*$_2$ still lingers (pestilence brought about by bad behavior: a punishment by the gods).
46. Nutton (forthcoming).
47. Breathing bad air as a form of causing common disease has a long history. "Malaria" was a term that was used to describe the bad effects of breathing night air (just the time that mosquitoes carrying the disease attack humans). For a literary account of this lingering idea, see Henry James' short story "Daisy Miller."

48. Lloyd (2007): 138.
49. Perhaps Straton as well? See Hermann Diels (1906): 101–112.
50. In the ancient world, this was particularly acute among the atomists who posited the void. Aristotle attacks this position in *Physics* IV, 6–9. In the seventeenth century, a similar argument was engaged in in letters between Henry More and René Descartes—for a discussion, see Boylan (1980).
51. To conceptualize this from a modern perspective, we cannot import ideas of either "air pressure" or "water pressure" because the idea of gas laws or hydraulic dynamics (that describe gas or fluid dynamics on the order of α-processes) were in no way present in the ancient world.
52. Of course, this is different in Aristotle where *dunamis* is contrasted with *energeia* and *entelechia* in a regular technical manner: potentiality, minimal actuality as functionally defined, and perfectly actualized. See Boylan (2009): 140–141.
53. The obvious objector would be the atomists who postulate at the most fundamental level: atoms, the void, and the πρόληψις (the swerve) that allows novelty.
54. For discussions on attraction in this context, see Iain M. Lonie (1981): 266–268; Michel Roussel (1983): 423–26; and Jacques Jouanna (1974): 351, n. 1. Supporting the context of a Hippocratic physiology, see Beate Gundert, (1992).
55. I am using a contemporary concept: clinical biomedicine, which refers to the interaction between patient and physician that depends heavily on the physician's own intuitively based dispositions from his own experience as opposed to medical researchers who have a different agenda—to figure out the laws of biology such that they can better create general principles that can serve all. This latter emphasis is behind the current trend in evidence-based medicine (which is supposed to trump the clinician's intuitively based dispositions). In the modern context, see Boylan and Brown (2002): ch. 7.
56. This backdrop position is what modern cosmologists in physics call the three possibilities: (1) the universe was caused and so there was a beginning; (2) the universe has been always and forever will be; and (3) the universe periodically comes to be and passes away. All three of these approaches have various advocates in the ancient world, for example, Plato, Aristotle, and Empedocles, respectively. Because there is a normative meta-story involved with each, there is a ground for natural normativity.
57. This argument persists in the contemporary era. I set out some of the principal current debates in my essay on health: Boylan (2014a).
58. . . . στάσις δὲ περὶ τοῦ ἐκκειμένου. Οἱ μὲν γὰρ εἶπον γίνεσθαι νόσους παρὰ τὰ περισσώματα τὰ γινόμενα ἀπο τῆς τροφῆς, οἱ δὲ παρὰ τὰ στοιχεῖα.
59. For a more complete discussion, see Laura Grams (2009).
60. This battle still occurs in the medical field as those who put forth general theories want to trump evidence-based medicine—see Boylan and Brown (2002).
61. *Fleshes*, I, 1–3, 8; VIII: 584–88.
62. *Ancient Medicine*, XXIV.
63. *Breaths*, I: 23–40. The author continues to suggest the methodology of treatment is to add or subtract from where the patient presents to the physician to get the patient into a natural condition.
64. Some of the reasons for this in our contemporary era focus upon differences in our body's structure, such as the length of our intestines—they can vary by as much as 57 centimeters—and the longer food is being processed, the more calories go to the body as opposed to waste; the sort of bacteria that reside

in our intestines—firmicutes are far more efficient in breaking down food and getting more calories out; and the amount of processing in the diet, since processed food is more readily broken up and utilized. These and other issues from the modern angle can be found in Novotny, Gebauer, and Baer (2012).

65. Beate Gundert (1992): 462–463, cf. *Vict.* (6.616:17; 618:3), *Salubr.* 77 (6.82: 21–84; 84: 6–7). Other causes for residue include the texture of tissue and the process of pregnancy.
66. XI: 44-XII: 1.
67. Πάντα δὲ ὁμοίως φησὶν Δέξιππος ὁ Κῶιος, ὃς οἴεται συνίστασθαι τὰς νόσους ἀπὸ τῶν τῆς τροφῆς περιττωμάτων, τοῦτ᾽ ἔστιν ἀπό τε χολῆς καὶ φλέγματος δυνάμεων γινομέςως περὶ μέρος καὶ περὶ ὅλον . . . (XII: 8–12).
68. Περι Νογσον Α, I: 1–30 (Littré, VI, 140).
69. Demont (1978): see his chart on page 378.
70. Demont (1978): 369–382.
71. Some excellent treatments include Giorgianni (2006), Joly (1970), and Lonie (1981).
72. Aeschylus: 272. G. Thompson (1941, rpt. 1972) 293–294; also J. Needham (1959): 25. I also treat these texts in Boylan (1984).
73. Euripides, (1909) vol. 3; Plato, (1902) vol. 4. For further discussion see: Anthony Preus 1970); and his book (1975): 52ff. See also Erna Lesky (1950): 18.
74. Cf. DK fragment 10, "How could hair come from that which is not hair or flesh from that which is not flesh?" The point is that flesh must send its "seeds" to the germ line to eventually produce flesh and so forth. The most recognized transit route would involve the movement of blood.
75. The second-class status of women in ancient Greece and Rome has been well documented. For a review of some of these arguments, see Mary Lefkowitz (2005) and Laura McClure (1999).
76. Examples of the moral power of women can be found in *Lysistrata* in the novel idea about ending the Peloponnesian War. Another example can be found in *Antigone* as the lead character sacrifices her safety so that her brother might have a burial that is demanded by natural law. Examples of the female warrior revolve around the tradition of the Amazon warriors. For an overview of this legend, see Lyn Webster Wilde (2000).
77. *Nature of the Child*, 492–496.
78. Galen, Kuhn (1964), XIX, 321.
79. Littré VIII: 470–785. I am influenced by Tage U. H. Ellinger (1952) and Iain Lonie (1981)fs. My translations follow from Paul Potter, *Generation* Loeb X (Cambridge, MA: Harvard University Press, 2012).
80. *Peri gones*: 474.
81. *Peri gones*: ch. 4.
82. *Peri gones*: 471.
83. Of course, this does not say much about the lower parts of the body. Since blood does not circulate back to the source (whatever that is said to be), this pangenesis theory cannot account for the legs and feet getting into the gonads.
84. According to my neurosurgeon daughter, Arianne, this conflict is evidenced in the modern general surgical approach within neurosurgery as opposed to orthopedic surgery. In the case of the former, a more empiric approach is adopted along the lines of being between "c" and "d" on the Figure 2.1 scale, while orthopedics is definitely at "a" since good practice in the orthopedic surgical subspecialty allows very little latitude for individuality.

3 Aristotle—The Origin and Function of Blood and Body Parts

How is it that Aristotle attempts to respond to the philosophy vs. clinical biomedicine debate? Well, frankly, by taking sides and trying to win the day for the philosophers. However, in the process he does create a systematic philosophy of biology that will set the stage for subsequent practitioners and theorists. Aristotle's work was widely influential in creating a coherent account that helped illumine some of the pivotal differences that will develop between the Dogmatists and the Methodists (see Chapter 4, this volume).

In this chapter, blood will be viewed first through a quick primer on Aristotle's philosophy of biology, then second to the role of blood in the classification of all animals, third to an exploration of the uniform parts, fourth to the creation of blood, and fifth to conception theory.

ARISTOTLE'S PHILOSOPHY OF BIOLOGY

Aristotle was not a physician, though his father was the court physician to Amyntas III of Macedon (father to Philip II and grandfather to Alexander), but Aristotle had a keen interest in biology.[1] It is my contention that this is the best way to understand Aristotle's philosophy.[2] Almost a third of Aristotle's writings can be classified as biology. Even works that are explicitly philosophical—*The Categories, On Interpretation, Physics, Nicomachean Ethics*, and *Metaphysics*—show the influence of the biological concepts set out in *The Parts of Animals* (Aristotle's key treatise on the philosophy of biology).[3]

This is not surprising. As mentioned in the first two chapters, many philosophers took an interest in biomedicine, and some physicians took an interest in philosophy. Those so inclined to philosophy are what I set out as leaning toward D-□. The attraction of mathematics and its certainty also turned some of these thinkers to astronomy as another form of natural philosophy.[4] Because biology is messier than astronomy, the D-□ crowd generally has to covert to D-◊ and its acceptance of *epi to polu* (for the most part). Within this context enters Aristotle, the Stagirite, who moved to Athens from Macedonia (a place hated in the eyes of the Athenians). He could never own land in Athens. He was denied the succession to Plato's academy largely because he was an accursed foreigner. Indirectly, being an outsider was also a benefit to Aristotle because the isolated thinker was able to set up a new approach to biomedicine along the lines of D-◊ and *phusis*$_1$ and $_2$.[5] He focused upon *types* and not *tokens*.[6]

There are really two Aristotles concerning scientific methodology. The first was pre-Plato's death: *The Posterior Analytics* (*APo.*).[7] It is here that he satisfied his teacher's inclination toward the mathematical. This work supported a quasi-deductive structure of scientific explanation, as opposed to the logic of discovery. In that book Aristotle is keen to address some key distinctions such as the role of *episteme* (exact knowledge), which is set out as that which cannot be otherwise—D-□.[8] This means that *episteme* must also be based upon principles that are true, primitive, and immediately grasped. These conditions logically precede the conclusion they are responsible for inferring. Thus, these principles are better known than the conclusion.[9] But they cannot account for themselves—which could lead to skepticism[10] (cf. II.19 for the secondary justification when this is not possible). However, *what is known* must derive from necessary principles.[11] Definitions must hold in every case—D-□ (73a28)—and be *kath' auto* (73a 34).[12] Aristotle proceeds to set out a number of other distinctions about the way things are known. What these formal principles seek is the knowledge of the reasoned fact (*dioti, APo.* II. 1, I.13). These distinctions follow because, after all, the book comes after the *Prior Analytics* where Aristotle invents formal logic. In the case of the *Posterior Analytics* we have one of the two versions of his considered philosophy of science that centers around answering the general questions: *hoti, dioti, ti esti*, and *ei esti* (exact apprehension of the fact, knowledge of the reasoned fact, how definitions and essences play a part, and the modality of the features of the discovery event via induction, *APo.* II.1). The *Posterior Analytics* provides a way to present data that one acquires. It seeks the most exact presentation possible: *apodeixis*, the conclusion of a syllogism. Knowledge of the conclusion of a syllogism is *episteme* (exact knowledge that meets the other requirements of *APo.*). The answers to the four questions in the presentational mode of the rest of the book represent the ideal that proponents of D-□ set out; however, this view of science really never guides the scientist—the other Aristotle.

Instead, we have another view of his philosophy of biology in the *Parts of Animals* I:

> In all study and investigation, be it humble or noble, there appears to be two types of proficiency. One is exact knowledge (*episteme*) of the [natural] subject; the other is a generalist's competence (*paideian*). For the latter has the educated person's power to judge what is put forth correctly and what is not. (*PA* 639a 1–7)[13]

Aristotle does not mean to denigrate or to exalt either sorts of proficiency. Both are necessary for natural investigations. The generalist's understanding is holistic and puts some area of study into a proper genus, while scientific knowledge deals with causes and definitions at the level of the species. These two skills are demonstrated by the following example:

> An example of what I mean is the question of whether one should take a single species and state its differentia independently, for example, *homo sapiens* (*anthropou*) nature [*phusis*$_1$] or the nature [*phusis*$_1$] of Lions or Oxen, etc., or should we first set down common attributes or a common character [to determine its nature, *phusis*$_2$]. (*PA* 639a 15–19)[14]

In other words, the methodology of the specialist would be to observe and catalogue each separate species by itself. The generalist, on the other hand, is drawn to making more global connections through an understanding of the common character of many species. Both skills are needed. Here and elsewhere Aristotle demonstrates the limitations of a single mode of discovery. We cannot simply set out a single path toward scientific investigation—whether it be demonstratively narrow (logical) exactness as per the specialist, or holistic understanding (the generalist's knowledge). Neither direction (specialist or generalist) is the one and only way to truth. Really, it is a little of both working in tandem (cf. *Categories* IV-IX). Sometimes one-half takes the lead and sometimes the other. The adoption of several methods is a cornerstone of Aristotelian pluralism, a methodological principle that characterizes much of Aristotle's work.

When discussing biological science, Aristotle presents the reader two directions: (1) the mode of discovery (genetic order) and (2) the presentation of a completed science (logical order). In the mode of discovery, the specialist sets out all the phenomena in as much detail as possible, while the generalist must use her intergeneric knowledge to sort out what may or may not be significant in the event taking place before her. This is because in the mode of discovery, the investigator is in the genetic order. Some possible errors that could be made in this order, for example, might be mistaking certain animal behaviors for an end for which they were not intended. For example, it is very easy to mistake mating behavior for aggressive territorial behavior. Since the generalist has seen many different types of animals, she may be in the best position (on the basis of generic analogy) to classify the sort of behavior in question.

In the mode of discovery one begins with the phenomenon and then seeks to create a causal explanation.[15] But how does one go about doing this? In the *Posterior Analytics* II.19, Aristotle suggests a process of induction that begins with the particular and then moves militarily to the universal. Arriving at the universal entails a comprehensive understanding of some phenomenon. For example, if one wanted to know whether fish sleep, one would first observe fish in their environment. If one of the behaviors of the fish meets the common understanding of sleep (such as being deadened to outside stimulus, showing little to no movement, and so forth), then one may move to the generalization that fish sleep.[16] But one cannot stop there. Once one has determined that fish sleep (via the inductive mode of discovery), it is now up to the researcher to ferret out the causes and reasons why in a systematic fashion. This second step is the mode of presentation. In this mode (as mentioned earlier—the ideal being the questions and distinctions set out in *APo.*),[17] the practitioner of biological science seeks to understand why the universal is as it is. Going back to the example of sleeping fish, the scientist would ask why fish need to sleep. Is it by analogy to humans and other animals that seem to gather strength through sleep? What ways might sleep be dangerous (say, by opening the individual fish to being eaten)? What do fish do to avoid this?

These and other questions require the practitioner to work back and forth with what has been set down in the mode of discovery for the purpose of providing an explanation. The most important tools for this exercise are the two modes of causal explanation: TE and ME.

For Aristotle there are four causes: material, efficient, formal, and final. The material cause is characterized as "[t]hat out-of-which something existing becomes" (*Phys.* 194b 24). The material has the potential for the range of final products. Within the material is, in a potential sense, that which is to be formed. Obviously, one piece of wood or metal has the potential to be many artifacts; yet the possibilities are not infinite. The material itself puts constraint upon what can be produced from it. One can execute designs in glass, for example, which could never be brought forth from brass.

The efficient cause is depicted as "that from whence comes the first principle of kinetic change or rest" (*Phys.* 194b 30). Aristotle gives the example of a male fathering a child as showing an efficient cause. The efficient cause is the trigger that starts a process moving. This sort of cause is what we think of cause today: the motions of bodies against each other that bring about an effect—such as a chisel against marble by the sculptor to ultimately create a human shape.

The formal cause constitutes the essence of something, while the final cause is the purpose of something. For example, Aristotle believed the tongue to be for the purpose of either talking or not. If the tongue was for the purpose of talking (final cause), then it had to be shaped in a certain way, wide and supple, so that it might form subtle differences in sound (formal cause). In this way the purpose of the tongue for speaking dovetails with the structural way it might be brought about.[18]

It is generally the case that Aristotle in his biological science interrelates the final and formal causes. For example, Aristotle says that the efficient cause may be inadequate to explain change. In the *On Generation and Corruption* 336a, Aristotle states that all natural, efficient causes are regulated by formal causes: "It is clear then that fire itself acts and is acted upon." What this means is that while the fire does act as efficient cause, the *manner* of this action is regulated by a formal/final cause. The formal cause (via the doctrine of natural place—that arranges an ascending hierarchy among the elements earth, water, air, and fire) dictates that fire is the highest level of the sublunar phenomena. Thus, its essence defines its purpose—namely, to travel upward toward its own natural place. In this way the formal and final causes act together to guide the actions of fire (efficient cause) to point upward toward its natural place.

Aristotle (at least in the biological works) invokes a strategy of redundant explanation. Taken at its simplest level, he gives four accounts of everything. However, in actual practice, it comes about that he really only offers two accounts. In the first account he presents a case for understanding an event via material/kinetic means (the material and efficient causes). For the sake of simplicity, let us call this the ME (materially based causal explanation) account.

In the second case he presents aspects of essence (formal cause) and purpose (final cause). These are presented together. For the sake of simplicity, let us call this the TE (teleologically based causal explanation) account. Now one way to understand these in the context of our philosophical/historical survey is to link ME to *phusis*$_1$ and TE to *phusis*$_2$. In the first instance, ME intends to give a purely material account—here supplemented with the motions of the matter moving in sequential explanation-type α fashion. In the second instance, TE extends to *phusis*$_2$ a new primary meaning: systemic purposefulness. The *Parts of Animals* and the *Generation of Animals* look at body parts in the context of what we now describe as physiology (as opposed to anatomy). Why must parts be shaped in this way or that? Such a question follows from *APo*. II.1 in beginning with the part (*hoti*) and moving to the account of its activity (*dioti*). But why must activity be purposive, that is, aim at something (the foundation of TE)? This is a deep-set orientation for Aristotle. The first sentence of both the *Nicomachean Ethics* and *The Metaphysics* points to such purposiveness: "Every *techne* and every *methodos* (and indeed every practical pursuit) seems to aim at some good; and it is rightly said that the Good is that to which all things aim" (*EN* 1094a 1) and "All humans, by nature, desire knowledge" (*Meta*. 980a 22). Such declarations stand in stark contrast to the Atomists and the Skeptics. For them, *phusis*$_3$ rules the day. Things occur *or not* by chance occurrences. There is no systemic purpose.

But what about the gods? Aristotle is not inclined to embrace the popular version of theism set out in his era. Instead, he has a rather deistic unmoved mover that is *technically* responsible for all movement in the universe—but it is neither a creator nor an active intervener in the physical events of the cosmos.[19] Thus, we are left with a new fashioning of *phusis*$_2$.

For an example of the dynamic between ME and TE, let us examine the process of respiration set out in *PA* I. Aristotle believes that material and efficient causes (ME) can give one account of the motions of the air in and out of the lungs for respiration. But this is only part of the story. One must also consider the purpose (TE) of respiration and how this essence affects the entire organism.[20] Thus, the combination of ME and TE are presented together as redundant explanations. The ME focuses upon how the nature of hot and cold air form a sort of current that brings in new air and exhales the old. The TE is about why we have respiration in the first place (to provide a source of external air to be transformed into internally charged air: *pneuma*).

In Aristotle's account of respiration, we are presented with a partner to TE and ME: *necessity*. When necessity attaches itself to ME, it is called simple or absolute (*haplos*) necessity. When necessity attaches itself to TE, it is called conditional necessity. Let us return to our example of respiration and examine these concepts in more detail:

> Here is an example of the method [I am advocating]: respiration takes place for a certain *purpose* [TE] and the stages of the process follow one another by [absolute] *necessity*.[ME] . . . sometimes necessity means that this or that is a purpose [conditional necessity] and that that these things *must* be so [in order to bring about a particular effect]; but sometimes it means [absolute necessity] for example, it is [absolutely] necessary that hot air goes out and comes in again when it meets resistance and the air must flow in [again]; this [process] follows by [absolute] necessity. (*PA* 642a 32–642b 3)[21]

This is a complicated passage. First, there is the (formal/final cause) TE of respiration. Respiration exists so that air might be brought into the body for the creation of *pneuma*.[22] If there were no respiration, there would be no intake of air and no way for it to be heated in the region of the heart and turned into *pneuma*—an element necessary for life among the blooded animals. Thus, the TE for respiration is for the sake of producing an essential raw material for the creation of *pneuma*.

The second mode of explanation, ME, concerns the material and efficient causes related to respiration. These have to do with the manner of a quasi-gas law theory. The hot air in the lungs will tend to stay there unless it is pushed out by the cold incoming air that hurries its exit.[23] (This is because 'hot' and 'cold' are two of the essential contraries hot/cold and wet/dry.) It is the material natures of the elements that dictate its motions. This is the realm of the ME.

ME is an important mode of explanation because it grounds the practitioner in the empirical facts so that he may not incline himself to offer mere *a priori* causal accounts. When one is forced to give material and kinetic accounts of some event, then one is grounded in the tangible dynamics of what is happening. This is one important requirement for knowledge of *phusis*$_1$.

It should also be noted that the ME account follows the general form of explanation-type α as set out in Chapter 2 in this volume. There is a sequence of events that regularly follow one another to effect some desired outcome: in this case, the entrance and exit of air from the lungs.

Now let's turn to necessity. Necessity can be represented as a modal operator (D-□ to D-◊ to E-◊) that can attach itself to either TE or ME. When it attaches itself to TE, the result is conditional necessity (D-◊ to E-◊). In conditional necessity, one must always begin with the purpose to be achieved. For example, if one assumes the teleological assumption of natural efficiency, "Nature does nothing in vain," [24] then the functions of various animal parts must be viewed within that frame. If we know that respiration is necessary for life, then what animal parts are necessary to allow respiration within different species? The acceptance of the end of respiration causes the investigator to account for how it can occur within a species. The same could be said for other given ends, such as "gaining nutrition," "defending one's self from attack," and "reproduction," among others. When the biomedical researcher begins his investigation with some end (whether in the mode of discovery or in the mode of scientific presentation), he is creating an account of conditional necessity.

The other sort of necessity is absolute (*haplos*) necessity (D-□ to D-◊) that is the result of matter following its nature, $phusis_1$ (such as fire moving to its natural place). The very composition of the material itself creates the dynamics—such as the quasi-gas law interactions between the hot and cold air in the lungs. These dynamics may be described without proximate reference to the purpose of the event. In this way, ME can function by itself along with simple necessity to give one complete account of an event.

In biomedical science Aristotle believes that conditional necessity is the most useful of the two necessities in discovery and explanation.[25] This is because, in biology, there is a sense that the entire explanation always requires the purpose to set out the boundaries of what is and what is not significant. However, in his practice it is most often the case that Aristotle employs two complete accounts of ME and TE in order to reveal different modes of explanation according to his doctrine of pluralism.

THE ROLE OF BLOOD IN THE CLASSIFICATION OF ALL ANIMALS

Blood is not confined to humans.[26] Many other animals have blood—such as viviparous quadrupeds (mammals) and oviparous animals (egg layers such as birds, fishes, reptiles, and amphibians). These roughly correspond to what (in the post-Linnaean classification) is called 'vertebrates.' It is of interest that Aristotle picked 'blood' to be the critical differentia. To appreciate this distinction, we must return to *APo.* and the weight pulled by 'primary' (the proper differentia).[27] Aristotle's theory of definition follows a class inclusion principle. We begin with a large class—such as the genus—and we add a differentia to get the species. One helpful device in coming to the proper definition is the convertibility (*antistrephein*) of the subject and predicate terms.[28] For example, "All humans are rational beings" can also be rewritten as "All rational beings are humans." This principle of convertibility corresponds to the modern logical understanding of coextensive sets. When there are two sets, α and β, and they have all and the same members, they are coextensive and share the property of identity: $\alpha = \beta$. However, convertibility is not enough. Aristotle also believes that only humans laugh (where laughter is a cognitive response to the disparity of two events' contranatural order).[29] Under this account it may be the case that "All humans are risible (capable of laughter)" and that "All risible beings are humans." Thus, the risibility is convertible, but it does not pick out the essence of humanity because it isn't primary. We are only able to laugh because we *understand* the incongruity with the natural order. Thus, *risibility* depends upon *rationality* for its very existence. Rationality is primary so that only rationality can set out the definition of what it means to be human.

But what about blood? What role does it play in the understanding of humans and other creatures? As mentioned earlier, having a red fluid that traverses the body is a large-scale dividing line in the animal kingdom according to Aristotle. From this fact we have to infer that there is some basic state of blood that is merely about transporting nourishment (*trophe*) to the body. Such a basic delivery device would be common to all the blooded animals: humans, mammals, oviparous quadrupeds, birds, and fishes. There may be other factors that are added to common blood, like specialized *pneuma*, that could account for variation in sensation and motor abilities. Animals, it should be remembered, differ from plants (another class of living thing) in their ability to sense and to move in specialized ways. This goes back to Aristotle's theory of soul (*psuche*) in which there are three souls: nutritive, sensitive/locomotive, and rational.[30] Plants possess the first only. Animals possess the first and the second. Humans possess all three. However, inasmuch as a human utilizes nutrition, then he has a physiological commonality to plants.[31] This would be understood on the table of predictables as making reference to "living thing."[32] Thus, it would make perfect sense that some animals and humans shared the common nutritional vehicle of blood—even though plants do not have blood. The expression of the nutritive soul in plants may be of a different variety—such as the clear, watery part of the blood (*ichor*—at least analogically?).

Then there are the animals that are characterized as bloodless.[33] These include *to entoma* (insected animals: Insects), *ta hostrakoderma* (shell-skinned animals: Testacea), *ta malakostraka* (soft-shelled animals: Crustacea), and *malakia* (soft-bodied animals: Cephalopods).[34] If these animals do not possess blood, and if blood is primarily the vehicle of providing nutrition to the body, then there must be an alternative: another fluid to provide nutrition.[35] The explanation in the first citation is that blood is treated as a basic nutritional conduit—like an irrigation ditch. 'Blood' in this context is the potentiality (*dunamis*) out of which the body becomes. In the second context, there is a very material (*$phusis_1$*) description:

> "As the nourishment (*trophe*) oozes through the blood-vessels and the passages in the several parts (just as water does when it stands in unbaked earthenware), flesh, or its counterpart, is formed: it is the cold which 'sets' the flesh and that is why fire dissolves it." (*GA* 43a 10–12, Peck, tr.)

I interpret these passages as asserting the fluid-nourishment function in humans. And since the nonblooded animals also have a nourishment need, there must be an analogous fluid that performs this function (*ichor*—correlate?).

Returning to humans, blood is instrumental to sensation and rationality. The powers of reason are generated in the region of the heart (associated by Aristotle as the producer of blood), but not necessarily in the heart itself. This is supported by three enigmatic passages in *peri psuche*:

> It is clear that the soul cannot be separated from the body, or at least certain parts are (if the soul has parts) . . . [But] there is nothing to prevent some parts from being separated because they are not actualities of any body part. It is also uncertain whether the soul viewed as a [controlling] actuality bears the same relationship to the body as a [controlling] sailor in a ship. (*De An.* 413a 7–9) . . . thus the soul which is called mind (*nous*) cannot reasonably be asserted to be blended with the body; if that were so it would acquire a quality such as hot or cold like the sensitive organs [especially touch]. (*De An.* 429a 25–27) . . . when mind [in the active sense] is freed from its present state it appears for what it really is: immortal and eternal. (*De An.* 430a 22–24)[36]

I take these passages to indicate that unlike *seeing* and the eye in which there is a clear unity of form and matter (an eye that cannot see is no longer properly thought of as an eye), thinking (the mind) may be separable from matter. In fact, unlike sight, it probably is not supported by any particular organ, as such. Instead, it is in the general region of the heart because the heart is the *arche* of the body, both in genesis and in actualized power:

> The first to be formed is the "principle," (*arche*) which in blooded animals is the heart and in others the counterpart of the heart, as I have said many times over. There can be no doubt about this, because our senses tell us that it is the first thing formed; but the truth of it is confirmed by what happens when the creature dies: the heart is the place where life fails last of all; and we find universally that what is the last to be formed is the first to fail, and the first to be formed is the last to fail. . . . In embryos, as soon as they are formed, the heart can be seen

> moving before any of the other parts, just like a living creature . . . the heart and the liver are visible in blooded animals as soon as they are formed at all, that is, when they are quite small: in eggs they are visible, just about the size of a point, sometimes as early as the third day, and very small ones are visible in aborted embryos. . . . With the common hen after three days and three nights there is the first indication of the embryo . . . where the primal element of the egg is situated . . . the heart appears, like a speck of blood in the white of the egg. (*GA* 741b 15ff; *PA* 666a 2–22, 665a 33–35; *HA* 561a 5, 9, 11)

There are a few quick items to note about this combined pastiche. The first is that Aristotle holds that within blooded animals the process of development is the same between hens and other blooded (though viviparous) animals. This is quite a leap. Second, he slips in a posit about the back side (death) that he intends to be implied by the evidence on the front side: the first to be formed is the last to fail. Even if Aristotle were correct about the first point (and in his sequential egg-cracking exercise seems to support this),[37] the evidence he puts forth does not support the second point (that the heart is the last to stop functioning). Though he does not give an empirical justification for this, his judgment has been historically influential. This priority of the heart as the central principle of life has guided the history of medicine until the 1960s when brain death took over as the principal indicator of death.[38]

THE UNIFORM BODY PARTS

Aristotle distinguished the body parts into two categories: the uniform parts (*ta homoiomere moria*) and the nonuniform parts (*ta anomoiomere moria*).[39] In the first category are blood, flesh, lard, suet, marrow, semen, bile, milk,[40] and the brain.[41] The uniform parts are, by nature, homogeneous.[42] They may be divided again and again without changing their character. The nonuniform parts include internal and external parts. Some external parts include the parts of the face, the whole face, the arm, hand, horns, and the like. Internal nonuniform parts are the various organs: heart, liver, spleen, lungs, etc.

Our keen interest here is in the uniform body parts—and most especially blood. Aristotle describes blood this way:

> Blood is always found in a container, viz., in the blood-vessels (*phlepsin*) and nowhere else except only in the heart. In no animal is the blood susceptible of sensation when touched, no more than the residue (*perittoma*) from the stomach is: the same holds true of the brain and the marrow. If you sever the flesh at any place, blood appears, provided the animal is alive and the flesh is not gangrened. If healthy, blood naturally has a sweet taste,[43]and a red color; if in inferior condition, either

> naturally or through disease, it tends to be black. Blood in its best condition is not excessively thick nor thin [except through disease]. . . . the blood in animals pulsates in the blood-vessels all over the body at once; and blood is the only fluid which remains throughout the whole body and throughout life so long as it lasts. . . . blood is formed by a process of concoction out of *ichor* . . . if the blood putrefies (*sepsis*) then it becomes pus (*puon*) and pus produces a chalk-like residue (*poros*). (*HA* 520b 12–22; 521a 6–9; 521a 18–21, translation after Peck)

What we see from this initial offering is the physical characteristics of blood and how it is observed in the body. It contains the power of pulsation within it (cf. Chapters 4 and 5, this volume). It is connected with life; it is omnipresent (since if you prick the skin anywhere in the body it comes forth); it is naturally red and unnaturally black (*phusis*$_1$); it is not too thick or thin;[44]and it comes out of *ichor*. This sense of *ichor* in our discussions in Chapters 1 and 2 in this volume would connect it with food (probably meat) that was part of the concoction process that created nourishment (*trophe*) out of food taken into the body. However, as we shall see later, it still retains some *phusis*$_2$ vital qualities.

Blood provides nourishment to the body[45] and plays a significant ME role[46] in the formation of other uniform parts.[47] Aristotle has blood produced in the heart (a nonuniform part), which holds blood "like water in a jar."[48] However, unlike a jar, ". . . Aristotle asserts that "it has three cavities, the largest being on the right side, the smallest on the left, and the medium-sized one in the middle" (*De hist. an.* 496a 19–21, Harris, tr.). This is an advancement in anatomy—though the exact nature of the third cavity has been the source of some controversy.[49]

From the heart the very warm blood rises to the brain, where it is cooled so that it might pass through very small vessels that are practically (though never completely) continuous with the skin where they are absorbed as nourishment. Aristotle sets out the composition of the blood as containing a watery part (*ichor*).[50] This watery part is cool and thin, connecting it with intelligence. The other parts of the blood are the fibers (*hines*),[51] which allow blood to coagulate.[52] The fibrous parts are hot and thick. Aristotle observes (from an external scan of visible blood vessels on the body—his own or others) that blood does not appear to be uniform in color. He extends this observation or inference to also include the consistency of the blood.[53] This assessment is probably an inference based upon differing concentrations of *ichor* and *hines* components. Other components—such as the various forms of internalized air—such as *pneuma*—do not affect the color or viscosity. It is also possible that there is the beginning of a homeostatic system set out: when it is *cold* outside, might the blood-making organ compensate by creating more *hines*, and when it is excessively *hot*, more *ichor*? There are no definitive passages, but the post-Alcmaeon's hypothesis of balance, and the various other Hippocratic Writers who supported similar positions, might render plausibility to such a conjecture.

However, there may be another component to Aristotle's analysis that is less empirical. Instead of engaging in the model of *phusis*$_1$ accounts in which the goal of the researcher is an explanation-type α outcome, Aristotle turns to various principles that are accepted via *a priori* principles. The result is an *a priori, phusis*$_2$ explanatory-type β account. For example, take the *a priori* principle that the *right* is nobler than the *left*.[54] Given that "hotter" is identified with life and "colder" with death, Aristotle combined these principles to conclude that the right side of the body is warmer than the left side.[55] This sort of reasoning is a bit like the presentation in Chapter 1 in this volume in which various authors utilize either *phusis*$_2$ or *phusis*$_3$ accounts based upon non-Empirical criteria to fill in gaps in the explanatory framework. In Aristotle's case here, he is committed to a cooking metaphor.[56] Blood is the result of various concoctions that are much like transformations that occur in cooking. The heat of fluids is crucial to cooking. Under this posit, there must be variable temperatures of blood. At the terminal point of concoction, the blood is at its hottest. Then it ascends and is cooled in the brain. As it descends via the large blood vessels, the right side gets its share first so that it remains at optimal temperature. What is left over goes to the left side so that it is cooler, and activities on the left side of the body are not as *arête* as those on the right side—after all, most people *are* right-handed, and there has been a long-standing prejudice among peoples in this region against the sinister left side.

This predilection is widespread. In *GA* 765b Aristotle contends that semen concocted on the right side is better than that from the left side.[57] Aristotle could have avoided this error by harkening back to the deductive and inductive rigor set out in the *Posterior Analytics*. Sadly, he did not.

A second problem with Aristotle's depiction of the composition of blood comes about when the "thin" and "thick" theories of blood are brought forward to account for intelligence. Thin blood is connected to intelligence, while thick blood is connected to strength.[58] If we apply this standard across the board, it would contradict Aristotle's doctrine about the superiority of males.[59] Peck correctly identifies the "thin" and "thick" theories of blood as a usage of a "Hippocratic" doctrine found in *On Regimen* I, 35.[60] In that treatise the argument is made that the coolest heat and the driest water are conducive of *phronesis*. This is because they are self-sufficient and able to exist at the extremes. The cool, wet is generalized to be beneficial for a number of medical needs because of its power of *phronesis* (intelligence and sensitivity). Aristotle used the same word and seems to be borrowing the connotation as well. However, several problems occur when Aristotle applies this to organisms. He wants to claim that it is this quality of thin, cool blood that distinguishes certain animals as being more intelligent than others.[61] This even applies to the analogous substance found in the bloodless animals.[62] Bees and ants, for example, have a more mobile faculty of sensation because of the thinness of their nonblood counterpart—further, they might be more intelligent than some of their blooded counterparts.[63]

Continuing on the theme of intelligence, we are faced with a great divide between those (beginning with Alcmaeon and followed by many of the Hippocratic Writers) who ascribed the brain as the seat of intelligence against those from the so-called "Italian school" who centered it around the region of the heart. This controversy is another key conflict between those who follow D-□ philosophy as opposed to the E-◊ crowd. Now Aristotle self-identifies as an Empiricist. However, the rest of human history declares his methodological identity to be a mixed bag, and by the time of Galen, he is called a Dogmatist (D-◊).

Argument 3.1: Blood and the Human Brain

Aristotle bases his argument that the brain is not responsible for thinking in the following argument from *Historia animalium*:

1. The brain has no blood vessels *in* it (only *around* it)—observation by Aristotle (presumably on either corpses or dead animals—more likely: *HA* 514a 15)
2. Blood only travels in blood vessels—observation
3. The brain has no blood in it—1, 2
4. Thin blood causes sensation, that is, is necessary for animal sensation (including reasoning)—conclusion from the *a priori* argument of (*PA* 656 b)

---

5. The brain has no sensation and is not the seat of reason—3,4

Premise #1 is difficult to understand. In almost every mammal, blood vessels are quite easily seen within the brain itself. Aristotle, however, believed that blood vessels only *surrounded* the brain in a delicate membrane:

> The brain itself in all animals is destitute of blood and no blood vessel great or small holds its course there. But of the remaining blood vessels that branch off from the last-mentioned blood vessel, some envelop the head, others close their courses in the organ of sense and at the roots of the teeth in blood vessels exceedingly fine and minute. (*HA* 514a 16–22)

Aristotle in this passage describes what we now call the meninges consisting in the dura mater, arachnoid mater, and pia mater. These do indeed form a net around the brain (hence the moniker 'arachnoid'). This required a detailed observation. However, to miss the anatomy of there being no other blood vessel in the brain itself suggests that Aristotle only looked at brains that had been dead for some time or perhaps one he ate at dinner.[64] However, if this were true, then Aristotle would not have seen the delicate membrane around the brain. This suggests that he has taken this observation from someone else—perhaps a butcher?

If Aristotle did make a firsthand observation, then his comparison to humans would be more accurate if it were a species closer to humans than, say, a fish (also classified by Aristotle as a blooded animal). Obviously, what Aristotle is searching for in his Argument 3.1 is the ME to support that blood only makes remote contact with the brain via an exterior net. This would allow the heated blood to be cooled so that it could descend to the rest of the body to provide nourishment (*trophe*).[65]

One wonders why Aristotle is so keen to deny that the brain is the seat of reason. One response might harken back to the earlier passages from *De Anima* that contrasted the loss of a particular function of the soul (such as sight) when the body part was impaired (i.e., the eye). This seems to be a cornerstone in Aristotle's *hulomorphism*. But if the brain were the seat of the rational soul, then this would entail that the rational soul, the seat of personal worldview, would also perish at death. One school of interpretation of Aristotle (that this author and most of the commentators in human history have adopted) says that Aristotle was keen to avoid this. Therefore, to put reason in the region of the heart but not beholding to any particular body part would be one way to avoid this outcome.[66]

THE CREATION OF BLOOD

We can begin our exposition on the creation of blood with the heart. Now Plato famously described the heart as a knot of blood vessels.[67] This is because Plato was still caught up in the D-□ goal of biological science. As Plato sets out in *Republic* IV, 439e-441b, there are three parts to the soul: *sophia* (wisdom), *thumos* (spiritedness), and *epithumos* (desire). Therefore, Plato in the *Timaeus* seeks to make his depiction of the body conform to these already determined characteristics. This is the paradigm of the "philosopher" position set out in Chapter 2 in this volume. It harkens after *a priori* over the empirical. The rational was intended to dominate the other two parts of the soul. This could lead to spiritual development between lives.[68] In order to accommodate this bifurcation, Plato situated reason in the head and emotions/desire in the heart. Thus, the head is the seat of the immortal part of the person. One reason for this is yet another instantiation of *phusis*$_2$: the theory of Forms. Since a sphere is a primal geometrical form and since ancient mathematics honored geometry as the most primary and perfect form,[69] it was here (the roughly spherical skull that houses the brain) that was most appropriate for reason to reside.[70] This use of *phusis*$_2$ was to be used in scientific justification up until and including Kepler.[71]

Harris (following Littré) contends that Aristotle was the first to describe the human heart as having more than one chamber.[72] In fact, it is the starting point for all blood vessels (*arche kai pege*). Aristotle asserts that the system of blood vessels is like a two-trunked tree that springs from a common root, the heart. From there are the two common blood vessels: the aorta and the vena cava whose origin is in the heart.[73]

When we examine Aristotle's theory on the creation of blood, we must set out several key terms that relate to the digestion of food and its transformation into nourishment for the body.

Pepsis (πέψσις). *Pepsis* is a very important concept in Aristotle's explanation of the digestive system. Roughly two things occur during *pepsis*: (1) food or other solid material is softened,[74] and (2) there is a separation (*diakrisis*)[75] of the food into component parts. The efficient cause of *pepsis* is heat. By means of heat the food is softened and broken up. The major results are nourishment (*trophe*) and a by-product (*perittoma*). This by-product is composed of solid and liquid elements, some of which are useful and some of which are useless. This reaction comprises the conjunction of matter (*hule*) and heat (*thermon*) to create even more heat,[76] along with nourishment (*trophe*) and the by-product (*perittoma*).

Pepsis causes a change (*metabole* or *alloiosis*) in the food. In common Greek usage, such heat alteration can include the ripening of fruit[77] and the cooking of food.[78] The author of *Ancient Medicine* asserts that in this process the substances undergo transformation due either to heat or the lack of it.[79]

Since this process requires significant alteration, the body requires several stages in which this alteration can occur. This is a harkening after explanation-type α.[80] I assert that there are four stages of *pepsis*: the first occurs in the stomach; the second occurs in the liver and spleen; the third occurs in the heart; and the fourth occurs in the gonads.[81]

Thermos (θερμός). *Thermos* is heat. There are two sorts of heat relevant to the human body: internal and external.[82] They correspond to the two types of cold. Both hot and cold are depicted as active agents. That is, the characteristics of wet/moist (*hugron*) and dry/hard (*xeron*) are determined by the active agency of hot and cold.[83] Of the two terms, hot is primary and cold is understood as relative to hot. *Wet* and *dry* are derivative characteristics that come to be by the hot and cold. When hot or cold *masters* (*kratounta*) some matter, the characteristics of moist or hard follow, by necessity, ME.[84] The process is best understood via the cooking metaphor.

Trophe, perittoma (τροφή, περίττομά). *Trophe*, or nourishment, is the principal end or purpose of *pepsis*. It is fluid in character. There are two types of *trophe*: (a) life sustaining (*threptikon*) and (b) growth promoting (*auxetikon*).[85] The former is necessary for living, and the latter is formed from what is left over. This nourishment constitutes only a small part of the food that has undergone various *pepses*.

What is *not* nourishment is residue (*perittoma*). It is the by-product of *pepsis*—the unconcocted.[86] The term 'residue' is used rather broadly to refer to both the useful and the useless substances. Among the useful functions of residue is the formation of male and female reproductive fluids: semen (*sperma*) and the female counterpart, menses (*katamenia*).[87] Lard, suet, and fat are also produced from residues,[88] as well as hair and nails.[89] Fat, hair, and so forth are healthy signs (in proper proportion)—they show that one has an adequate amount of food, for in the process of digestion they are provided only after all other uses of residue have been fulfilled. The useless residue can include unusable parts of food as well as colliquescence (*suntegma*), a separation from growth-producing material.[90] This *suntegma* can roam the body, causing all sorts of mischief from pus and boils to tumors and possibly even gonorrhea.[91]

Since Aristotle believed the heart to be the first organ produced via his examination of developing fertilized chicken eggs, he concluded that it must be the source of blood because blood is necessary for the nourishment and development of the body. Thus, blood's purpose (TE) is sustaining the uniform (*homoiomere*) and nonuniform (*anomoiomere*) parts. Since this has to do with nutrition and since the heart is depicted as the final stage of the digestive process,[92] it is appropriate that blood, as a final product, is produced here. The TE of the heart is to perform the last concoction of digested materials into its rarefied form: blood. The ME that brings this about is that (1) the incoming nourishment is sufficiently prepared, that is, it contains as little residue as possible; (2) there is intense heat; and (3) there is the pneumatization of the blood. This ME is provided by Aristotle through the motions of the heart:

> There are three movements which take place in the region of the heart, which seem to be of the same character, but really are not: palpitation, pulsation, and respiration. Palpitation is a violent concentration of the heat due to the chilling effect of waste products or secretions . . . the heart pulsation is due to the expansion by the heart of the liquid food-product which continually enters it . . . respiration occurs owing to the increase of the hot substance which contains the nutritive principle. Because just as all the other parts need food, so does this, and even more than the other parts; for it is the cause of nourishment for the rest. Now as it increases it must cause the organ to rise. One may regard the structure of the organ as very like that of both the bellows in a forge. (*De resp*. 479b 19–22, 480a 3, 480a 17–19)

Palpitation occurs from residue within the heart and is not important to this ME except as an explanation of why some metabolic processes are less efficient than others. Pulsation (*sphugmos*) results from the influx of liquid nourishment from the liver and spleen. This satisfies the first systematic requirement.

The second is controlled by respiration. The lungs are said to act mechanically, like a bellows filling up with air.[93] The action is controlled by the heart, which uses the air as a cooling device.[94] The heart expands owing to the *pepsis* and new *pneuma* being formed from the heart.[95] One must assume that the reaction proceeds rapidly and gives off heat. But where does the heat come from? If heat and nourishment lead to pneumatized blood, then how does this heat originate? There are mechanisms to control it, but what is the source? There is no systematic discussion of this point except to say that it is an ongoing process.[96] This is how blood is created as the outcome of the third *pepsis* that begins with digestion and ends up in the heart that creates the final product.

CONCEPTION THEORY

We can continue this examination of the creation of blood in the heart. After blood has been perfected, it exits the heart via blood vessels. There are two parts to the blood: nourishment and residue. The nourishment gets used up as it delivers its sustaining qualities to the uniform and nonuniform parts. Thus, the TE for blood as nourishment is to grow and to sustain the body.

The TE for residue is chiefly for reproduction (what is left over can sustain other functions mentioned earlier). The ME is more complicated.[97] The chief problem is the transformation of the residue into the male *sperma* and the female *katamenia* (the menses, which is the analogue to the male seed). How does a material that in both males and females is roughly the same—the residue from the heart—become transformed into two different reproductive substances?

Well, let's begin with the differences between male and female blood:

> The blood (*haima*) of females differs from that of males. Thus, if both are equally healthy and of the same age, the blood in females is thicker and blacker . . . blood (*katamenia*) is more plentiful in women . . . in very young animals blood (*haima*) is more *ichor*-like. In the aged it becomes thick, black, and scanty. (*HA* 521a 21–23, 26, 33)[98]

The male blood is more balanced and redder; therefore, it is better than the *thicker* and *blacker* female blood. Thus, our answer to the two reproductive substances begins with the nature of blood. Second, we must consider the qualities of the residue in the gonads: "The *sperma*, it is clear, is from a bloodlike substance that was formed from the residue of the nourishment."[99] This is also true of *katamenia*.[100] It is important to note that the residue that makes up these generative substances is different from other residues formed from the stomach, liver, and spleen. None of these other residues is bloodlike.

Aristotle noted that of the two generative substances, *katamenia* was most like the residue described and so must undergo the smallest alteration. In fact, the manner in which the uterus acts on this residue is somewhat unclear:

> When these are overfull of nourishment (which owing to its own coldness the female system is unable to concoct), it passes through these

> extremely fine blood-vessels into the uterus (*GA* 738a 13–14) . . . While it is within the mother, then, it develops slowly on account of its coldness, since development is a sort of concoction, and if a thing is hotter its concoction is easy (775a 18) . . . The reason why the uterus is always inside is that it is the container for the young creature while it is being formed, and this needs protection, shelter, and concoction (719a 30) . . . The male and the female are distinguished by a certain ability and inability. Male is that which is able to concoct (765b 15).

In some of these passages it appears that the uterus concocts the residue creating *katamenia*, while in others it seems as if this residue itself is *katamenia*; the textual evidence is unclear.[101] What is clear is that whatever goes on in the uterus is different from what goes on in the testes. Either there is no *pepsis* in the uterus while there is a fourth *pepsis* in testes, or there is a quasi-fourth *pepsis* in the uterus and a more complete fourth *pepsis* in the testes. My guess is that Aristotle thought there might be some changes taking place in the uterus, but that these changes did not constitute a full-blown fourth *pepsis* (a 3.5 *pepsis*?). However, it should also be noted that Aristotle's discussion of *katamenia* is more extended and detailed than that of his predecessors.

Male *sperma* is clearly concocted yet another time in the testes.[102] The ME of this *pepsis* is unique. There is no by-product. All the material given to it is used. The *pepsis* is not a separating of various elements, but is a process that unites and compacts them. What Aristotle is trying to account for is the viscosity of semen. The cohesion of semen, Aristotle would probably say, comes about through *optesis* (external, dry heat).[103] The process yields a dry exterior membrane. However, the most important part of the ME is the potency of the *sperma*. This is dependent upon the compaction (*sunestos*) and the presence of *pneuma*.[104] When either of these factors is missing to any appreciable degree, the male is infertile.[105]

In the testes, there is the final true *pepsis*. There are some passages that suggest that the *katamenia* may also undergo a fourth *pepsis*,[106] but most passages seem to reject this notion (cf. *GA* 719a). When enough semen has been produced, any extra residue that comes in its direction is sent back to the body to create hair and nails. The whole system depends upon heat to create *pepsis*. The hotter each *pepsis*, the better and more efficient is the process. Each department meets its needs and there is ample left over. However, as one gets older or when one sickens, the heat is decreased and the least essential areas are excluded in a reverse hierarchy triage.[107]

The end point of the process in the male seed, *sperma*, is how it acts as the efficient cause in reproduction (*GA* 716a 5). But what does this mean? Obviously, that the potentiality of the female *katamenia* is acted upon by the male *sperma*. This initial dynamic is termed *synesis*. It refers to something like the explanation-type α. It is sequential in the fashioning of the developing *kuema*, which can include everything from the zygote to the mature fetus.[108] From the first, the *kuema* is viewed as a separate entity developing epigenetically within a given environment.[109] The mother and father cease to have a proximate relation to the *kuema*. It is from this point (though it may contain potentialities from its parents) that its development is materially caused by its own parts.[110] The relative places of *sperma* and *katamenia* in reproduction are not relevant at this stage, for it is the heart that is immediately formed and, in turn, forms the blood and the other uniform parts, which then fashion the nonuniform parts (the various organs of the body). This describes an epigenetic process that is contrary to the preformation development account that attaches to the pangenesis account of many of the Hippocratic Writers. Under pangenesis, various potentialities are released from each part of the body via the blood to the male and female gonads where they are ready to be fully assembled in what we moderns call the zygote-blastula developmental cycle. From there, the little, fully assembled entity simply gets bigger. In contrast to this, Aristotle's epigenetic account has a presentation process in which the dynamic powers of the *sperma* are activated via the fourth *pepsis*. Once the *sperma* makes contact with the *katamenia*, then a sequential process of development comes to be, one after the other (*tode meta tode*), but not by the preceding part (*oux hup ekeinou*).[111] The parts come from blood, but blood itself does not fashion them as a sculptor fashions a statue—this is the role of *sperma*, which fashions the parts in a sequential order because of a logical order inherent in nature (*phusis*$_1$ and *phusis*$_2$).[112]

> One can grasp from these examples how the male contributes to generation. Not every male emits *sperma* and those who do, this *sperma* generates no part of the *kuema*. . . . In the same way nature [*phusis*$_1$ and *phusis*$_2$] acting in the male in those [species of animals] which eject *sperma*, uses the *sperma* as a tool, which has actual movements just as the movement of tools in a craft produces things. This is because in some way the motions of the craft is in them. (*GA* 730b 8–11, 19–23)

The artist in this example is *phusis*$_1$ and *phusis*$_2$. The latter uses the means of the former to bring about its TE intention via ME. The tool is the *sperma*. The motions of the tools are regulated by the craftsman in order to fulfill the TE-directed mission. In the ME, the *sperma* travels to the uterus (where Aristotle says conception occurs) and activates (*sunista*) the *katamenia*.[113] This analogy is repeated a number of times.[114] It seems evident that Aristotle did intend TE (*phusis*$_2$—Nature) to actively use *phusis*$_1$ in carrying out the ME. The craftsman metaphor does just this.

The 729a 10 passage, which states that the male furnishes the form and source of movement, is somewhat misleading. Even though the male does furnish (*parexetai*) both, only one comes from the male qua male: the efficient cause. Since no kinetic cause may be expressed *simpliciter* (it must always be restrained by some principles of TE), the efficient cause supplied by the male brings about the potentialities of form. But these do not actually come *from* the male, but from nature (*phusis*$_2$ and *phusis*$_1$) working through the *sperma*, which controls the motions and directs them to their given end. Consequently, although the motions from the male are controlled to initiate the "form" of the *kuema*, this form is not, strictly speaking, from the male qua male. The particular male in question (the token) is merely the instrument of *physis*$_2$ (the type) which is driven by *type-directed* outcomes. The example that Aristotle gives just after the 729a 10 passage is of rennet curdling milk. This explanation-type α process is regulated via *phusis*$_2$ via TE.

All of this might mitigate the seeming bias that the human male (as token and type) is completely domineering in a comprehensive way.[115] Now in the *Politics* Aristotle does say that:

> The male is by nature (*phusis*$_2$) superior and the female inferior; the male is the ruler and the female is the subject. . . . The slave has not got deliberative reason at all and the female has got it, but without full authority. (*Pol.* 1254b 13, 1260a 13)

This passage is supported in the *Generation of Animals*:

> The male qua male is active and the female qua female is passive (*GA* 729b 11). . . . The male provides both the form and the source of movement and the female provides the body and the matter (*GA* 729a 10). . . . a woman is like an infertile male (*GA* 728a 18) . . . a woman is like a deformed male (*GA* 737a 28). . . . The proximate motive cause, insofar as it subsists within the *logos* and the form is better and more divine than the matter (*GA* 732a 3–4).[116]

These passages seem to indicate that the normal or natural (*phusis*$_2$) result of copulation is a male child.[117] However, Aristotle cannot ignore the fact that some children take after their mother. The response is that females are "regular" defects that occur. All defects come about through (1) the relapse principle by which things are said to turn over to their opposite, or (2) the principle of unmastered or unruly matter.[118] Only the former can affect the sex of the child. "So that if the [*sperma*] movement gains mastery then the [*kuema*] will be made male, if not it will be female and a male which is like the father and not like the mother" (*GA* 767b 21).[119] The inheritance principle is separate from the sex-determining principle—though the distinction is small.[120] Thus, the relapsing can refer to resemblance to the mother or distant relatives, or it can refer to the sex-determining principle.[121] Presumably there is a separate trait for each. But if the female comes about via a *regular* process, how can Aristotle be justified in calling it a deformity? The very word 'deformity' indicates that something has gone "away from" the proper or normal form. When such an action occurs 50% of the time or more, it seems odd to call it a deformity. Perhaps Aristotle would say that it occurred through this process less often (that is, "normally" through this process a male was born) because there was a secondary factor that interfered with nature (*phusis*$_1$): environmental causes.[122] With the relapse factor working alongside environmental causes, Aristotle thinks he can account for the great number of females who are born.[123]

Under this account, female contribution to the fetus is only an instance of a sort of unnatural occurrence. This is supported by Aristotle's contention that females do not contribute to the soul (*psuche*) of their offspring. What is important is that, taken as it is, the female has no actual *psuche* in her reproductive fluid—but then neither does the male.[124] The explanation is that neither can have *psuche* in their reproductive fluid, since if this were possible, there would be two souls within one individual.[125] Rather, the form or proximate cause for the *kuema's* soul comes again from *phusis*$_2$ via the *phusis*$_1$ agency of the male. Presumably, the only *psuche* that Aristotle cared about for humans is the rational soul. This is because it is the defining differentia for people. Thus, if agency of the male brought about the rational soul, it would make him more divine.[126]

Perhaps the two other souls (nutritive and sensory/motor) might be provided jointly or by the female only? This is only a conjecture because Aristotle is silent on this question. This is a shame because it could have offered Aristotle a natural way (*kata phusis*) to include the female in conception. His deformity-centered account of the female contribution seems contrary to his own standard of critical Empiricism, and causes him to lean toward the D-□ philosophical accounts against those of either critical Empiricism (D-◊) or clinical Empiricism (E-◊). It is a methodological choice that is inconsistent with his biomedical theory set out at the beginning of this chapter.

NOTES

1. Nutton (2013) describes how the "bio" and the "medicine" come together, 119–122.
2. See Boylan (2010), cf. Marjorie Grene (1963). Another sort of example along these lines is Sarah Francis' (2011) article that argues that Aristotle's account of *akrasia* in the *Nicomachean Ethics* has a root in his understanding of biological physiology.
3. Boylan (1983): chs. 2 and 3.
4. Dreyer (1906) chs. 5–7; Mueller (1981): 185–187.
5. Of course, it is interesting to examine just how Aristotle thought about nature (*phusis*). For a discussion of this, see Boylan (1983): 69, 87, 89–109,126–128; Boylan (1985); and Boylan (1986b).
6. For a fictional account of Aristotle in Athens, see Boylan (2010).
7. As I noted in Chapter 2, n. 14, this volume, Aristotle does not pick up his project of the presentation of a science set out in *APo.* in the biological works. This is particularly true in his use of definitions; cf. Pellegrin (1985), 99. Gotthelf (2012), 178 tries to mount a defense of Aristotle by appealing to partial definitions. For a middle position, see Devin Henry (2008).
8. *APo.* 71b 9.

9. *APo.* 72a 25.
10. This is the stance of *phusis*$_3$.
11. *APo.* 73a 21.
12. These can concern static or causal relationships. 'Human' defined as a rational animal would be an example of a static relationship. It is about class inclusion. However, 'eclipse' is a term with causal, theoretical elements that requires causation along with class inclusion in creating its definition. See Devin Henry (2008).
13. My translations in this chapter are very influenced by the Loeb translators with some notable exceptions. I would encourage readers to compare the two together.
14. The Greek text (as per the translation) only mentions *phusis* once, and yet I give several different understandings of *phusis* as per Chapter 1 in this volume. This is because the passage seems to assume different implied sensibilities of *phusis* in its exposition. This is why this passage is so controversial among scholars of Aristotle's philosophy of biology, cf. Boylan (1983), Balme (1972), and Gotthelf (2012).
15. *PA* 646a 25.
16. *On Sleeping and Waking* 455b 8, cf. *On Dreams* 458b 9.
17. Though this is an ideal, it is not an ideal that ever seems to interest Aristotle in the biological works. This puts me alongside Barnes and Pellegrin.
18. *PA* 660a 27–32.
19. *De Caelo* 288a 22–27. Aristotle's account of god is set in terms of motion. Motion means that one moves from potentiality to actuality over some background condition—such as place (locomotion). Now things that are in motion are put into that state by virtue of some other entity acting upon them. This causes a regress to an original cause. The original cause must be something that moves without moving itself (or else the regress would continue). This is the origin of the unmoved mover *Phys* VII, 1; VII, 4–5. This unmoved mover is a deistic depiction of God. How this detached source of ME and TE puts events into natural motion (*De Caelo* III.2 301b 17, cf. *Phys.* II 192b 20) is a source of some dispute. It will be the position of this book to interpret "nature as the cause of motion" to include in its meaning a reference to the unmoved mover: God—as in the following "οὐθὲν γὰρ ὡς ἔτυχε ποιεῖ ἡ φύσις" (*De Caelo* 290a 32–33, cf. *PA* 658a 9—Galen uses this phrase as well: *Peri Spermatos* I.5). From the epistemological side, there are two important arguments from the *Metaphysics*. In the first (982a 5–983a 20), Aristotle asserts that natural wonder makes humans search for first principles and primary causes. In the second (1071a 35–1074b 20), this searching seeks a highest cause. Plato's theory of forms is rejected in favor of Aristotle's own answer to the searching for *arche* and *stoikeia* and this is eternal substance. Instead of a separate, parallel realm, why not bring the unmoved mover into a monomaterialist realm instead: what historians of philosophy call the deist track.
20. *PA* 642a 31–642b 4.
21. The entire passage is as follows: Δεικτέον δ' οὕτως, οἷον ὅτι ἔστι μὲν ἡ ἀ ναπνοὴ τουδὶ χάριν, τοῦτο δὲ γίνεται διὰ τάδε ἐξ ἀνάγκης. ἡ δ' ἀνάγκη ότε μὲν σημαίνει ὅτι εἰ ἐκεῖνο ἔσται τὸ οὗ ἕνεκα, ταῦτα ἀνάγκη ἐστιν οὕτως ἔχειν, ότὲ δ' ὅτι ἔστιν οὕτως ἔχοντα καὶ πεφυκότα. τὸ θερμὸν γὰρ ἀναγκαῖον ἐξιέναι καὶ πάλιν εἰσιέναι ἀντικροῦον, τὸν δ' ἀέρα εἰσρεῖν. τοῦτο δ' ἤδη ἀντικροῦον ἐστιν, τοῦ ἐντὸς δὲ θερμοῦ ἀντικόπτοντος ἐν τῇ ψύξει τοῦ θύραθεν ἀέρος ἡ εἴσοδος καὶ ἡ ἔξοδος. ὁ μὲν οὖν τρόπος οὗτος ὁ τῆς μεθόδου, καὶ περὶ ὧν δεῖ λαβεῖν τὰς αἰτίας, ταῦτα καὶ ταιαῦτά ἐστιν. (P.A. 642a 31–642b 4). Abraham Bos (2002) holds this as a *pneuma*-making process that can account for the "instrumental body."
22. See the discussion in Chapter 2 in this volume on *pneuma*.

23. cf. *On Breath* 481b 11.
24. *GA* 741b 5, cf. 739b20, et al.
25. *PA* 639b 25. I should note that John Cooper (1987) and David Balme (1987) demur on their being any connection to material components and their nature. They want to reconfigure necessity as exclusively a province of teleology. However, Richard Sorabji (1980: 155–156) quotes Wolfgang Kullmann (1974) on a reading of *PA* 640a 33-b 1 as putting the emphasis upon the material in an interaction with teleology (my ME and TE). My reading sides with Sorabji and Kullmann.
26. It is of interest to note that Aristotle holds that the differentia for the genus animals that picks out humans is 'reason.' This reason resides in the region of the heart and so (one might infer) is connected in some way with blood. If this is correct, then blood may vary between species. However, a counter argument to this would be for reason to exist in the region of the heart, but with no ME account (the material blood). Instead, on this alternative reading reason may not be materially based. Such an account relies heavily upon the troublesome passage in *Peri Psuche* on the immortality of the soul (because reason may not be materially situated): see note 34.
27. *APo*. 83b 1, cf. 78a 28–78b 4.
28. *APo*. 78a6–14, cf. *APr*. II. 2–4 and *Top*. I.15, 167b 1–3.
29. *Rh*. 1370b 29–1372a 3.
30. *De An*. 413a 20–413b 2 ff.
31. Of course, this is a big mistake in biology since plants produce nicotinamide adenine dinucleotide phosphate (NADPH) and adenosine triphosphate (ATP) molecules via either photosynthesis or via the Calvin cycle, while animals (for the most part) utilize the Krebs cycle to create ATP molecules. Since plants have no analogous parts to animals, one might assume that they created nourishment in a different manner than animals. However, Aristotle did not write a treatise on plants, leaving that to his student, Theophrastus.
32. *Categories* 1a 20–1b 24; 2a 13–27.
33. The main determinant for "blood" is the color red for an interior fluid substance. On this account Aristotle is correct. However, if "blood" is defined as an interior fluid substance that delivers various needed nourishment to the body, then Aristotle is wrong. The so-called bloodless animals have such a fluid, but because they do not possess hemes, there is no red color. It is hard to fault Aristotle here. Perhaps he could have done more in analogical reasoning of which he is so fond.
34. Here I am following Peck and Forster (1968): 23.
35. cf. *PA* 668b 15 and *GA* 743a 10.
36. The context of these passages is as follows: ὅτι μὲν οὖν οὐκ ἔσιτν ἡ ψυχὴ χωριστὴ τοῦ σώματος, ἢ μέρη τινὰ αὐτῆς, εἰ μεριστὴ πέφυκεν, οὐκ ἄδηλον ἐνίων γὰρ ἡ ἐντελέχεια τῶν μερῶν ἐστιν αὐτῶν. οὐ μὴν ἀλλ᾽ ἔνιά γε οὐθὲν κωλύει, διὰ τὸ μηθενὸς εἶναι σώματος ἐντελεχείας. ἔτι δὲ ἄδηλον εἰ οὕτως ἐντελέχεια τοῦ σώματος ἡ ψυχὴ ὥσπερ πλωτὴρ πλοίον. (*De An*. 413a 4–9). διὸ οὐδὲ μεμῖχθαι εὔλογον αὐτὸν τῷ σώματι ποιός τις γὰρ ἂν γίγνοιτο, ψυχρὸς ἢ θερμός, ἢ κἂν ὄργανόν τι εἴη, ὥσπερ τῷ αἰσθητικῷ νῦν δ᾽ οὐθέν ἐστιν. (*De An*. 429a 24–27); ἡ δὲ κατὰ δύναμιν χρόνῳ προτέρα ἐν τῷ ἑνι, ὅλος δὲ οὐ χρόνῳ ἀλλ᾽ οὐχ ὁτὲ μὲν νοεῖ ὁτε δ᾽ οὐ νοεῖ. χωρισθεὶς δ᾽ ἐστὶ μόνον τοῦθ᾽ ὅπερ ἐστί, καὶ τοῦτο μόνον ἀθάνατον καὶ ἀιδον (οὐ μνημονεύομεν δέ, ὅτι τοῦτο μὲν ἀπαθές, ὁ δὲ παθητικὸς νοῦς φθαρτός), καὶ ἄνευ τουτου οὐθὲν νοεῖ. (*De An*. 430a 21–25).
37. Today we hold that in chickens, it is the spinal cord that is formed first, but this is not visible unless the researcher uses staining techniques that were not available to Aristotle. For the naïve Empiricist point of view, Aristotle is correct.

38. Aristotle's depiction of the heart as the clear indicator of death follows from its being the *arche* of the body. This standard held fast until the later twentieth century. For a discussion of this change, see Margaret Lock (2001), Gary Belkin (2014), and C. U. M. Smith (2013).
39. The uniform parts precede the non-uniform parts in genesis and constitute the ME for the non-uniform parts' coming-to-be *PA* 647b 10, 640b 20, 647b 22ff.
40. *PA* 647b 13.
41. *PA* 652a 25.
42. *PA* 647b 18.
43. Presumably, this observation is made from eating cooked meat and the juice thereby produced. There are no other accounts of Aristotle consuming blood from live humans, for example. He was not "Dracutotle."
44. This seems to be a nod to Aristotle's penchant for trusting the median, as per his ethics. It is unclear which is the more primitive insight. If it is the ethics, then Aristotle here is engaging in $phusis_2$.
45. *PA* 647b 27.
46. Cf. *PA* 651a 12.
47. *GA*. 726b 5.
48. *PA*. 650b 10.
49. Littré and Harris support this claim, but identifying which of the three chambers we are talking about is complicated by his description of the various blood vessels: "All of them, even the two small ones, have a connection with the lung, and this is quite clearly visible in respect to one of them. Below, at the place of attachment from or to the largest cavity, there is a connection to the great blood vessel, and from the middle cavity there is a connection to the aorta" *De hist. an.* 496a 22–27 (Peck, tr.). It is assumed that Aristotle is talking about some cavity between the left and right ventricle. But is this either of the atria or some small heart valve space that Aristotle is observing (see Vesalius VI, 12)? Arthur Platt (1921) noted Galen's skepticism to this in Galen, *De Ven. et Art. Diss.* 9 Kühn II, 817. Another key factor that is discussed by Thomas Huxley (1880): 1–5 and William Ogle (1882): 523 is that the problem may be a prolapse caused by the way Aristotle prepared his specimens—by strangulation. Harris also discusses this at 125–130.
50. *PA* 651a.
51. *HA* 3, 6.
52. *Ichor* and *hines* appear at *Timaeus* 83c 5 and 85c 3 and d 2, respectively. In this context, "serum" is identified as the rare, innocent part of the blood, whereas "fibers" are said to (1) maintain the balance between the rare and the dense, and (2) congeal the blood. Obviously, Aristotle picked up on the "rareness" of *serum* from the previous ambiguities on *ichor* (common pus vs. blood of the gods; see Chapters 1 and 2 in this volume). But concerning *fibers* he has adopted the second usage and ignored the first. In fact, he has made the 'fibers' to be the dense part when Plato conveyed this function to phlegm.
53. *PA* 647b 35.
54. *PA* 647a 10. Likewise, the *front* is to be preferred to the *back* and the *upper* over the *lower*. On the front-to-back, see the growth patterns of hair at *PA* 658a 23 and the heart's position at *PA* 665b 18. That the upper is "on top" over the lower, see *PA* 665b 20 and 686a 13.
55. *PA* 667a, cf. 648a.
56. Alberto Jori (2005) makes this case as well that Aristotle depicted the creation of blood like a cooking machine that has the heart as the starting point.
57. In *De An* 405b 28, Empedocles' use of *a priori* principles of natural direction to explain plant growth is criticized by Aristotle. But Aristotle here is doing the same thing—see his use of "right" over "left" to explain optimal

(male-oriented) semen concoction (*GA* 766b 35). This is a holdover from the various usages of *phusis* set out in Chapter 1 in this volume.

58. *PA* 648a 5.
59. It should be remembered that Aristotle is a systematic philosopher (as opposed to an eristic philosopher, like Plato—see Chapter 5 in this volume on this distinction). This means that he intends for all his distinctions to be applicable across his opus. But it is difficult to square his poor treatment of women with his doctrine that coolness in blood is associated with intelligence. Women are associated with the cool and moist and men with the hot and dry. This appears to be a systemic problem for Aristotle because it would infer that women are more intelligent than men—a position that Aristotle does not hold.
60. See Peck's comments in his translation of *The Parts of Animals*: 648a.
61. *PA* 650b 20.
62. Of course, insects have blood. It is just not blood with hemoglobin. The circulatory system in insects is an open gut system where the blood flows about in the thorax cavity. Aristotle noted that there was an analogous fluid, but because it was not red and the organisms did not possess vessels of passage, he did not call it blood.
63. *PA* 650b 24.
64. Lones (1912: 179) suggests that Aristotle only saw a cooked brain. My daughter, Arianne, who is a neurosurgeon, tells me that when dealing with a live human brain (under surgery) or even a human brain shortly after death that the issue with there being blood vessels in the brain is undeniable to the naïve Empiricist. This would indicate that Aristotle did not view newly killed warriors in a medical setting. But surely he could have seen animals in a butcher's shop (the agora). Since Aristotle is keen on the analogy of comparative anatomy, he could not have failed to realize that brains are bloody—and not just the meninges.
65. *PA* 652b 21–25, cf. 653a 5. It is interesting to note that the brain *does* act as a cooling mechanism in certain animals such as the cat, sheep, dog, and certain simians. This is accomplished by the blood traveling through the carotid *rete* before going through the circle of Willis. It is within this network that the warmer arterial blood loses heat to the cooler venous blood. It is interesting to note that this basic structure was known anatomically, though not physiologically, as early as Aristotle but was expanded by Galen and possibly Herophilus. For a discussion of this phenomenon, see M. A. Baker (1972). Preus (1975) also makes an interesting comment on the possibility of there being a uniform correct temperature for the body. If such a temperature existed, there would be no need for a cooling mechanism—except as a homeostasis system. However, outside influences such as climate and food ingestion (Preus, 122) affect temperature (cf. *Airs, Waters, and Places*).
66. There are, of course, a few who demur on whether Aristotle had an interest in an immortality of the soul or even if Aristotle actually believed in a god (see Balme 1972, 87 and Gotthelf 2012, 173–175). This is because these authors were keen to remake Aristotle in their own image. There is a long history of this in the commentator tradition. But because some moderns are atheists themselves, they seek to interpret their hero in the same mode—just like Gilbert Ryle (1949), who wanted to assert that Plato had no Theory of Forms because that would have committed Plato to accept some form of theism. Rather, I take the standard view to be that of Ross (1923, 179–186) who took Aristotle to be a forerunner of deism. Others who have weighed in on this issue include Abraham Bos (2003) who believes (by examining texts by Iamblichus and Augustine) that Aristotle could be interpreted as a dualist. In another paper (2002), Bos holds that *pneuma* in the blood forms a kind of "instrumental body" that could support the continuation of the soul.

67. *Timaeus* 70b. Levin (2014) argues that Plato has more to say about medicine than is generally acknowledged. She cites comparisons to various Hippocratic Writers.
68. See *Phaedrus* 246a-257b. In the charioteer story, the two horses that guide the chariot are the noble (*kalos*) and the ignoble (*aiskros*). This foray through the Forms was important for the transmigrated soul to improve.
69. Árpád Szabó, part 3.
70. *Timaeus* 44d.
71. Dreyer, ch. 15 and Kuhn (1957): 209–219.
72. Harris, 122. However, he did not have the chambers very accurate: "As I have already said, it [the heart] has three cavities, the largest being on the right side, the smallest on the left, and the medium-sized one in the middle. All of them, even the two small ones have a connection with the lung, and this is quite clearly visible in respect of one of them . . . there is a connection to the great blood vessel [vena cava] and from the middle cavity there is a connection to the aorta" (*De hist. an.* 469a 4–19, tr. Harris, my brackets).
73. *PA* 654a 32-b 13.
74. *Mete.* 379b 18, *GA* 775a 17, *Pr.* 907a 18.
75. *Mete.* 340b 3.
76. Adding heat to a reaction that creates even more heat is an important concept in the history of science: the exothermic reaction. Of course, it would be incorrect to characterize Aristotle's exposition in this way since the term 'exothermic' is embedded within a modern understanding of chemistry. However, it is an interesting detail to note in the history of chemical reactions.
77. *Mete.* 380a 11; Theophrastus *HP* 1.12.2, 5.1.3.
78. *Mete.* 380b 13, 381a 23.
79. *Ancient Medicine* (XVIII, 10–20).
80. cf. Galen, *Nat. Fac.* 3.4.
81. Boylan 1982.
82. The external can be hot essentially (as in firewood) or accidentally (as in boiling water). This and other distinctions made in *PA* 648b 10 are of less interest to the present discussion. For an analysis of heat as a power, see C. Strang (1961).
83. *Mete.* 378b 20.
84. *Mete.* 379a.
85. *GA* 744b 34.
86. *GA* 745b 18.
87. *GA* 737a 3.
88. *PA* 651a 20.
89. *GA* 745b 15ff.
90. *GA* 724b 26.
91. David Balme in his Clarendon translation of *De generatione animalium*, p. 147 makes this suggestion. He cites Michael of Ephesus 47.3 and Galen K XIX, 426 to support his claim. Compare also to the E-◊ authors discussed in Chapter 2 in this volume and their account of residue as the cause of disease.
92. *De juv.* 469a 3–4; see: Boylan (1982): 109–113.
93. This comparison to the bellows, if taken seriously, also accounts for heat. But this contradicts the lungs' function as a cooling instrument, which is emphasized in *De resp.* 475a 25 and *De juv.* 469b 8.
94. *De resp.* 480b. Aristotle does not mention the mixing of the air and blood, which might have served as a source of *pneuma*. However, in *HA* 495b 10 and 496a 30, an alternative path from the trachea to the heart for *pneuma* is vaguely described.
95. *De resp.* 479b 32.

96. *De resp.* 480a 5.
97. For a more thorough description of these, see Boylan (1984).
98. *Ichor* here is related to youth—presumably meaning vitality. This would represent a *phusis*$_2$ power of vitality—perhaps related to the Chapter 1 reference in this volume to *Iliad* V. 340 as the blood of the gods. At *HA* 489a 24 *ichor* is compared to the watery part of the blood (τὸ ὑδατῶδες τοῦ αἵματος, *PA* 651a 17). Whether this is a positive note related to Aristotle's inclination to thin blood and cognitive capacity or whether this watery part is just underconcocted residue is impossible to judge with certainty. Following Demont and Jouanna (1981), I am inclined to the more exalted view in this context—especially since *ichor* is said to be that out of which blood becomes—at least part of the ME, but possibly part of the TE, as well.
99. *GA* 726a 10.
100. *GA* 727a, 727b 30, 750b 11, 751a 4.
101. In *HA* X there is an extended discussion of the role of the uterus in reproduction giving some weight to the *katamenia* being part of a dual seed theory (though the *katamenia* is of an inherently lower quality since female blood is of a lower quality—but greater quantity—than male blood), cf. *GA* 725b 3, 728a 26, 728b 22, 737a 28, 750b 4, 766b 14, and *PA* 689a 12.
102. *GA* 765b 5, 719b 2.
103. *Optesis* dry heat (uniformly dry exterior and interior heat) is to be contrasted to *hepsesis* wet heat (uniformly dry exterior and wet interior).
104. *GA* 765b 4; 2.3.
105. Aristotle suggests a test for male fertility of placing semen in a tub of water. If it floats, the semen is not concocted enough, and the male is infertile; *GA* 747a 5.
106. cf. *GA* 727a, 727b 30, 750b 11, 751a 4.
107. *GA* 746b 25ff.
108. *GA* 728b 34.
109. *GA* I.8; 746a 20.
110. *GA* 734a 20 ff.
111. *GA* 734a 28.
112. This is not a perfect explanation-type α account, but supposedly there are potentialities for various parts within the blood, which express themselves sequentially. Cf. Galen's comment at K IV 529 on Aristotle's method as well as his own theory, K IV 660.
113. It is not clear whether Aristotle thought that the uterus "sucked in" the *sperma*. Galen quotes Hippocrates against uterine "sucking in," K IV 524. Aristotle is rather vague: *GA* 739b 5.
114. *GA* 723b 28, 731a, 738b 25, 740b 25, 743a 25, and at *PA* 641a, 652b 15.
115. This intuition on my part is disputed by Robert Mayhew, ch. 7.
116. Mayhew (2004) suggests another way to look at Aristotle's view of women as deformed men. He suggests that the castrated male is a better model to understand the difference: 54–68.
117. Balme (1980), 2 supports this interpretation; however, Jori (2005) believes this can be mitigated if we admit the existence of "male" and "female" blood. Presumably, this would require a different *pepsis* mechanism in each.
118. *GA* 768a 15 and 770a 6ff.
119. This is a controversial passage. Montgomery Furth (1988), 141 sees any maternal movements as a collapse of the form (male)-matter (female) dichotomy. Johannes Morsink (1982), 138, 171 tries to maintain the consistency by using the *epi to polu* tack. This means that female resemblance is contrary to nature. John Cooper (1988), 18–19 suggests that female movements are irrelevant and the traditional presentation is unchanged. Finally,

Devin Henry (2006), 269–274 argues that whether the mother contributes movements in the process of development is irrelevant. What is relevant is whether these movements have any effect. Henry thinks they do via an analysis that is consistent with the type-token distinction.

120. *GA* 768a 21–22.
121. *GA* 766a 16-b16, 768a 14 vs. 768a 31, 768b 8, 768b 15.
122. *GA* 766b 35. These environmental causes might be excess coldness that does not allow enough heat for the male to fully concoct. Since females are said to be cooler, excess coldness in the outside air could create female offspring. This might be the cause of the ancient idea that if a woman goes out in the night and exposes her pregnant abdomen to the cool night air that a female might result.
123. There is yet another sort of deformity that comes about via weak seed or troublesome *suntegma* and yields gross deformity such as "child headed" calves. These deformities are clearly *para phusis*$_2$. In these cases there is no TE at work since "from chance nothing comes about with purpose" *APo*. 95a9.
124. *GA* 736b 8, 737a 17.
125. *de An*. 413b 17.
126. *GA* 732a 3–4.

4 Hellenistic, Alexandrian, and Roman Methodological Battles

The historical transition from the Hippocratic Writers and Aristotle to Galen has been depicted in various ways. This book's emphasis has been to examine treatments of blood through the lens of various senses of *phusis*. Using the methodology of twenty-first-century philosophical logic and philosophy of science, the intent has been to show how these writers make important contributions to the timeless conundrums that face all philosophers of science. For our purposes in this volume, the next writers to discuss are Diocles of Carystus and Praxagoras as setting out an initial program that will set the stage for third-century BC biomedical writers.[1] On the one hand, both of these writers fit in with the Critical Empiricism program of Aristotle that tended toward D-◊.[2] These writers' acceptance of a realistic $phusis_1$ context that could be discovered and described caused others in the Hellenistic period to depict this group of writers as Dogmatists (*dogmatikoi*) or Logical/Rationally based Physicians (*logikoi*).[3] Both epithets have something to recommend them. This volume will use the term "Dogmatist," though what is most important in this nominal baptism is that according to our chart of modality set out in Chapter 2 of this volume, being *logical* means that one is probably situated at D-◊ with a desire to be at D-□ (only those nasty individualistic details get in the way).[4] One of the reasons that it is difficult to pinpoint Dogmatists as exactly as the Empiricists or Methodists is because the Dogmatists were a less coherently recognizable group than either the Empiricists or the Methodists. This accounts for the Dogmatist span from D-□ to D-◊.

The Dogmatist approach is logically troublesome to skeptically inclined practitioners who wish to value experience over theory, and they have been termed Empiricists.[5] Finally, in a move to simplify the process and focus upon the disease itself in a streamlined context are the Methodists who value signs and symptoms over causes.[6] The Empiricists incline toward the clinical biomedical practice, as per Chapter 2 in this volume, while the Methodists will be situated between the Empiricists and Dogmatists.

As always, our brief exploration of these three directions of biomedical practice will focus upon an examination of the underlying philosophy of each—especially as it affects blood, the heart, and blood vessels. The overall purpose is to view the way various practitioners made conclusions on medical theory and medical practice on the basis of what they believe biomedicine to be.

THE DOGMATISTS

Let us begin this section of the book with some consideration of Diocles and Praxagoras. First, let us situate these authors historically. "Diocles of Carystus, indeed, and Praxagoras of Cos, the son of Nicarchus, . . . both lived shortly after Hippocrates."[7] In fact, Diocles of Carystus is often called the younger Hippocrates.[8] Diocles is also connected to Aristotelian doctrine and that of Aristotle's pupil, Theophrastus.[9] Galen claims that Diocles was the first to write on anatomy using animal dissections on mules to make conjectures about the human female uterus.[10] It is difficult to get an exact dating on whether Diocles was a contemporary or just followed Aristotle.[11] Besides writing on anatomy Diocles wrote a number of other works, including *Gynecology, Bandaging, Treatments, The Surgery*, and *Affection, Causes and Treatment*. However, because no single treatise exists in whole, we are forced to do with quoted fragments and testimonials.[12]

Like Aristotle, Diocles is said to have claimed a leading role for the heart over the brain. He was aware that the heart had chambers. The exact number is a subject for debate, but there were at least two and possibly four[13]—set on the principle of bilateral symmetry and based upon the analogy of the brain having two lobes. He claimed that the left ventricle was the core of innate heat and innate *pneuma*.[14] Wellman believed that this was a doctrine acquired from Empedocles and that this trove of innate *pneuma* allowed Diocles to give *phusis*$_1$[15] support to the conjecture that thinking originated in the region of the heart rather than the brain.[16] This also situates Diocles with the Hippocratic Writers associated with Cnidos/Sicily (if such a distinction really holds).[17]

Diocles held that *pneuma* flowed in the blood vessels undifferentiated. This follows Aristotle's general description.[18] Blood was one of the four humors, and disease occurred because of excess or deficiency of one or more of the humors.[19] What *pneuma* added to blood was a particular refinement or power.[20] Because *pneuma* was so important, the blockage of it was also a significant cause of disease. This follows the general direction of the so-called Sicilian/Cnidian writers who viewed blockages (generally to do with the flow of blood or a constituent part being blocked by phlegm) or putrefaction (generally to do with digestion and its role in creating blood) as significant causes of disease. Blockages can be caused by the temperature of the blood. The extremes were bad: hot (via bile) caused the blood to boil and curdle, and cold (via phlegm) caused it to solidify—like ice.[21] These blockages can affect the body as well as the mind.[22]

Since *pneuma* is created from air and since *pneuma* is mixed into the blood, it is important to account for the input of air. Aside from the respiration via the nose that goes to the lungs directly via the trachea, Diocles held that the pores in the skin (which account for sweat and can be seen by the naked eye) are also involved in respiration.[23] However, it is not entirely clear that pore respiration is expelled via the lungs or via the pores.[24] This process was described by Plato and probably by Philistion. In the physiological account that would allow these two processes to occur, it seems that a distinction between the "thick" artery and the "thick" vein was made in a rough fashion. This would correspond to identifying the difference between the vena cava and the aorta.[25] This is a first step in distinguishing the arteries and the veins.

Diocles was considered an expert in diet and exercise as a therapeutic.[26] Along with respiration and the creation of *pneuma*, innate heat was a causal consequence (*phusis*$_1$). This allowed for a theoretical basis of the various causes of diseases cited earlier.[27] He is considered to be a Dogmatist because he sought underlying universal causes for biomedical phenomena.[28]

Praxagoras, a "descendent of Asclepius,"[29]was from Cos. His father, Nicharchus, was also an eminent physician.[30] I will take Praxagoras to be slightly younger than Diocles.[31] Praxagoras (highly influenced by Aristotle) was known as a follower of the Dogmatic or Logical school.[32] He is linked by Anonymous Londinensis to Hippocrates,[33] particularly in the doctrine of bubbles in the blood arising from the feet that can bring about epilepsy by causing blockages.[34] The material connection is clear. Bubbles rise. If they start in the feet, they can rise up the legs to the trunk, arms, and head. Through common experience one can observe that bubbles alter fluid dynamics. The application to the body is materially based, *phusis*$_1$. This sort of general, material account is characteristic of the Dogmatists.

Praxagoras was also known as a teacher of a number of prominent physicians: Herophilus of Chalcedon (a founder of the Alexandrian school),[35] Phylotimus,[36] Plistonicus,[37] and Xenophon of Cos.[38] Measured on the basis of one's pupils, Praxagoras rates rather highly in this period of history.[39]

Praxagoras identified ten to eleven humors. This comes from Rufus of Ephesus.[40] But blood is not on the list. Galen interpreted these as kinds of phlegm, the black and yellow bile.[41] The reason for this taxonomy is to support a theory of digestion (and putrefaction—according to his student Plistonicus[42]) that could account for imbalances and the resulting disease. When one asserts a greater number of animal juices (humors or *chumes*) there is immediately a more nuanced, logical approach (the more unique *tokens* there are, the more resultant *type connections* that can be linked to them—see discussion in Chapter 2, this volume). This multiplies the ontological universe of disease types. Everything becomes more complicated in a theoretical sense, but the poor sick guy is still in the same amount of pain. Those inclined to a theoretical (Logical/Dogmatic) grounding of medicine into a general theory of biomedicine will favor more distinctions and unique tokens over fewer. Thus, this part of Praxagoras' theory puts him foursquare into the Dogmatist camp.[43]

However, there is a strong Empirical strain to his writing as well. This follows along the line of Aristotle's critical Empiricism (see Chapter 3, this volume). Aristotle's influence is also seen in Praxagoras' defending the Stagirite's position on the heart as the center of reason. Praxagoras does this by claiming anatomical proof that the brain was the mere outgrowth of the spinal cord.[44] The interest in anatomy as a cornerstone of theory sets Aristotle and Praxagoras apart as advocates of *phusis*$_1$, which requires connection to a material account. This is what classifies the Dogmatists apart from earlier thinkers who often grounded their abstract accounts in a general acceptance of supernatural principles, *phusis*$_2$ (see Chapter 1, this volume, on the various uses of *phusis*).

This inclination to base theory upon empirically examined anatomy (albeit animal anatomy) led Praxagoras to his central accomplishment in the area of blood: the distinction of arteries (*arteriai*) and veins (*phlebes*). Until Praxagoras, the blood vessels, *phlebes*, were all lumped together. It was Praxagoras who distinguished them by an aspect of their material character.[45] He distinguished arteries from common blood vessels on the basis of their origin:[46]

> The largest artery has its origin in the left hollow of the heart like a trunk of all the arteries in the animal. The artery winding around the heart rises first from this artery and has been accordingly named [aorta] by those who study anatomy (tr. Steckerl, 1958).[47]

With the arteries and veins distinguished, it was next incumbent for Praxagoras to give the reason why (cf. *dioti*, Aristotle, *A.Po.* II.i). The answer is that they performed different functions. The arteries carried *pneuma* only, while the veins carried blood mixed with some *pneuma* (cf. the "bubbles" mentioned earlier).[48] This account prompts further queries about the "why" for *pneuma* in the body. The answer is twofold: (1) the role of *pneuma* as generally necessary for bodily heat (remember when a person dies, he becomes cold), and (2) the role of *pneuma* in creating a nervous system by which bodily movement had a material account.[49]

First is the role of *pneuma* in bodily heat. Praxagoras believed that bodily heat was created by the consumption of foodstuffs in the process of digestion (cf. the writing of his pupil, Phylotimus).[50] This led to "hunger cures" for fevers.[51] But this account is limited because, following Aristotle, digestion went through its final *pepsis* in the liver.[52] For Praxagoras, this was the origin of blood. This account would therefore say how blood is heated, but what of *pneuma*? Is it also produced in the digestion process as nourishment, *trophe*? Steckerl (1958) opines that there are actually two sources of *pneuma*: one found in the digestion process, a source from heat, and a second source from respiration.[53] We will deal with respiration momentarily, but for now the question is if air is a by-product of digestion—perhaps on the evidence of burping and flatulence—then *how* does the air get down to the feet so that it might form bubbles and rise? It would seem that the rising bubble theory works best if all the bubble roiling begins in the gut (the source of the freed or processed air). But then it could not be a cause for problems occurring in the lower half of the body since the mechanics require that bubbles rise. This is an unresolved problem. But at least Praxagoras' digestion account of *pneuma* seems to solve the question of bodily heat, and it does (though controversially) get *pneuma* into the blood.[54]

Pneuma is part of a process of respiration. It begins in the lungs (and possibly the skin pores) and ends up in the heart (the source of the arterial system that carries only *pneuma*). But how does the *pneuma* get concocted? And what is its purpose? If the heart is a center whereby ordinary air is transformed into vital *pneuma*, then how does it happen? This is a question appropriate to a Dogmatist.

> What important benefit is provided for us by respiration? Is it the creation of the soul itself, as Asclepiades says? Or not the creation, but some kind of nourishment, as Praxagoras, the son of Nicarchus says? Or a certain cooling of the innate heat, as Philiston and Diocles held? (Steckerl, tr., 1958)[55]

This is a suggestive passage because it seems to support Steckerl's (1958) dual nourishment hypothesis: *pneuma* is formed in two ways: (1) through digestion in the liver, and (2) through respiration (presumably beginning in the lungs generally and then moving to the small arteries in the lungs, back to the heart where it is condensed into a sort of steam, *atmodes*). This steam, presumably, is also a source of heat since in cooking steam only arises after a fluid is heated. It is unclear whether these two sorts of heat account for all bodily heat (excluding innate heat) or whether they merely supplement innate heat. There is not enough textual evidence to support resolution of these questions.

Second is the role of *pneuma* in creating a nervous system by which bodily movement has a material account. According to Praxagoras, the *pneuma*-carrying arteries become finer and finer until their hollow tube construction collapses and the result is a nerve, which is the proximate deliverer of pneumatic impulses to the body. *Pneuma* may also be a cause of the pulse. Galen attributes three arterial properties to Praxagoras: "palpitation" (*palmos*), "tremor" (*tromos*), and "spasm" (*spasmos*).[56] These three words all indicate movement so that they might be associated with the movement of the artery.[57] However, what seems to be lacking in these terms is the rhythmic, regular movement. It seems that Praxagoras believed that the pulse was connected to various events occurring in the arteries themselves (presumably via the regular activity of the *pneuma* in the artery or endothelium) and not from any motions of the heart.[58]

The activity of the *pneuma* in the artery that will minimize into nerves is a quasi-prototype of the modern nervous system. It gives a reason why the arteries only carry *pneuma* and why the arteries become nerves. Nerves refer to small and large sinews in the body—often connected by predecessors to ligaments and tendons (the material strings that can move joints and account for human motion). As prescient as this might seem to modern readers, Praxagoras' account of a nervous system that began in the heart (with *atmodes*-charged *pneuma*) and then extended to the rest of the body was criticized by Galen:

> Aristotle and Praxagoras assert . . . that the heart is the origin of the nerves . . . that many of their other anatomic observations are correct, you can learn from their extant writings; but that they were either themselves completely blind or conversed with blind men when they wrote on the origin of the nerves. . . . [Praxagoras] liked to argue against Hippocrates to exclude completely the brain as the origin of nerves . . . then he got very bold in putting forward this lie that the arteries dividing up in their course become narrow and turn into nerves. The arteries had thus a nerve-like, though hollow form. Through the process of their being repeatedly split in the animal, the lumen becomes so small that the walls fall together. When this first happens, the vessel appears already a nerve (tr. based on Steckerl, 1958).[59]

The essence of Galen's criticism really revolves around the Cnidian cum Aristotelian notion of the heart as the seat of the soul (at least rational and sensory/motive). There is *not* an objection of the dynamic qualities of *pneuma*. Also, the mechanics of Praxagoras' account seem not to be so much of a problem had the entire enterprise had a different origin/principle, *arche*. Thus, we can move on with the assurance that various key principles have been enunciated: the difference between arteries and veins, the role of *pneuma* as a more material dynamic mover, and the discovery that the pulse can become a diagnostic tool because of its relation to what was later called vital *pneuma*. On the whole, Praxagoras made a significant step forward in our philosophical appreciation of blood vessels, physiology, and the pulse (a useful diagnostic device).

Aristotle's theory of critical Empiricism advocated examining the anatomy of various animals in order to get a grasp on underlying causes. Thus, our next two biomedical writers in the Dogmatist tradition that we will briefly examine are the anatomists Herophilus and Erasistratus.[60] Herophilus and Erasistratus were younger than Praxagoras,[61] and are contemporaries of sorts. Celsus says:

> Hippocrates of Cos was a man most worthy to be remembered for his professional skill and eloquence . . . after him was Diocles of Carystus, next Praxagoras and Chrysippus, then Herophilus and Erasistratus (tr. Spencer).[62]

Celsus continues to speculate on two ways that physicians investigated disease: the evident and the hidden.[63] The evident refers to what we moderns call symptoms (see the Empiricists later). These are clearly identifiable conditions, such as sensations of hot and cold in the patient, which is determined either by patient report or by the physician laying his hands upon the patient. The hidden refers to physiological theory (including anatomy)—enter Herophilus and Erasistratus:

> [Herophilus and Erasistratus] believe it impossible for one who is ignorant in the origin of diseases to learn how to treat them suitably . . . they inquire why our blood-vessels now subside, now swell up. . . Moreover, as pains, and also various kinds of diseases arise in the more internal parts, they hold that no one can apply remedies for these who is ignorant about the parts themselves; hence, it becomes necessary to lay open the bodies of the dead and to scrutinize their viscera and intestines. They hold that Herophilus and Erasistratus did this in the best way by far, when they laid open men whilst alive—criminals received out of prison from the kings—and whilst these were still breathing, observed parts beforehand nature had concealed, their position, color, shape, size, arrangement, hardness, softness, smoothness, relation, processes and depressions of each, and whether any part is inserted into or is received by another (Celsus, tr. Spencer).[64]

There are several points to be gained from this passage. First, that Herophilus and Erasistratus engaged in dissection of corpses because they wanted to investigate "the hidden." This is one of the key calling cards of the Dogmatists.[65] It is tied to the epistemological "appearance versus reality" problem.[66] One camp says that there is more there than can be seen without further investigation. This leads some to posit explanations that are, in principle, hidden to naïve Empiricism (what is true is only that which can be verified by the naked eye under ordinary circumstances). Critical Empiricism (à la Aristotle) suggested expanding the circumstances of inspection (such as animal dissection) and the creation of scientific laws using the rules of logic and unverifiable general principles about biological systems—such as "Nature does nothing in vain." Along this line, Herophilus' and Erasistratus' engaging in dissecting of corpses ups the ante by a measure of kind: instead of dissecting pigs, dogs, and cattle to understand humans, they have broken the long-standing prohibition about not only *touching the dead*, but also cutting them apart. This was a big step in the general Greek worldview, which paid great attention to the treatment of the dead.[67] Surely, if one wished to understand *the hidden*, then the hidden must be made empirically evident. The easiest way to do this is by dissecting a human corpse. Corpses do not display movement on either the exterior or the interior. This is the limitation on the knowledge gained via dissecting a human corpse. On the upside, the question of anatomy, given that it is performed before substantial material decay, can show clear, important relationships between body parts. What it cannot show is physiology: how the various organs operate in a functional, living organism.

Vivisection can solve all of that—in theory. For our purposes, vivisection will refer to the cutting open of a human for the purpose of anatomical and/

or physiological knowledge. It is not a therapeutic procedure. Thus, vivisection gives nothing valuable to the patient. It is solely for the sake of scientific knowledge.

In our modern context we do not perform vivisection. Instead, we engage in surgery, which is justified by its therapeutic purpose. If added information about anatomy or physiology is gained in the process, then this is a fortunate side benefit.[68] Certainly, from the patient's point of view, there is a world of difference between being cut open in the modern operating theater under the modern protocols that have been advanced and being cut open in Alexandria in ancient times without these protocols.[69] Under modern conditions (the patient is anesthetized and there are various monitors on basic bodily functions with options available should blood pressure or some other vital sign change dramatically); the cutting open of a human can be practiced safely and humanely. However, without these safeguards, the cutting open of a human creates significant pain, and this can put the individual into shock so that only very gross facts of human gut functioning can be observed. But sometimes the gross facts can be important. For example, William Harvey was known to use vivisection on dogs regularly in his presentation meant to show the heart to be a pump. He would cut up the dogs and show how the heart rhythmically pumped blood out of the body by severing an artery (and drenching the first rows of his observers).[70] In the case of Herophilus and Erasistratus, it is probable that dissections of cadavers yielded more pertinent information than vivisection.

It is thought that Herophilus and Erasistratus did their anatomical research in Alexandria (a city founded by the conqueror, Alexander of Macedon) and then his possible resting place after his premature death.[71] This gave a certain prestige to the city. Ptolemy I had made it a global center of learning that included both the premier library of everything that had been written and a liberal environment for further research as they created an advanced research center: the "Hall of Muses" near the royal palace.[72] With this government support, anatomical research in the Western tradition began here—though its subsequent progress is shrouded in mystery since the old prohibitions still held elsewhere and continued until the sixteenth century—though Nutton has suggested to me that it may be even earlier: the fourteenth century via process of autopsy).

It is also appropriate that Egypt be the chosen country for a Greek city. This is because of the long tradition of medicine within Egypt.[73] Herodotus even declared that Egyptians were among the healthiest in the world (next to the Libyans) due to their large number of physicians.[74] One common aspect in interaction between the Greek medical tradition with the Egyptian is the role of magic (*phusis*$_2$).[75] For example, von Staden cites various incantations that accompanied therapy. For the cessation of bleeding, one incantation is to be recited from a written document (papyrus) and then inserted into the anus of the patient:[76]

> You should flow out, mucus, son of mucus, who breaks the bones, who destroys the brain, who chops around in the bone marrow, who causes

> the seven openings in the heads of the followers of Re to become ill, and they turn to Thoth in prayer. . . Come out on to the earth: putrefy, putrefy, putrefy, putrefy, putrefy (Papyrus Ebers 763, tr. von Staden).[77]

As we examined in Chapter 1 in this volume, particularly in the context of *ichor* there was a strong undercurrent of *phusis*$_2$ within the emerging discourse on blood. This magical undercurrent continued in Greek-speaking regions on their own, without help from the Egyptians. During the Hellenistic period there are many religious healing centers in which various spiritual devices are employed, including purifying baths, the laying on of hands, incubation sleep (in special chambers—sometimes with nonpoisonous snakes roaming about), curative waters from sacred fonts, dream analysis with priests, and sometimes even dramatic *katharsis*.[78] The existence of such centers of healing that are slanted toward magic indicate a Greek tradition that embraced *phusis*$_2$ regardless of any interaction with Egypt.

In fact, Herophilus makes it very clear that he is not inclined toward *phusis*$_2$ explanations by the fact that he does not regularly make any reference to the gods as he describes anatomy and his view on the practice of medicine. The one exception is when talking about pharmacology he says that they are in the "hands of gods."[79] However, this is most likely a general metaphysical, commonplace reference.

It is not a mere coincidence that Herophilus and Erasistratus made strides in pharmacology within the context of Greek medicine[80] while in Alexandria.[81] Some have contended that this is because the Egyptians had formalized precise measurements for creating potions—as opposed to the looser qualitative style of the Greeks.[82]

Within this rich Alexandrian context Herophilus, Erasistratus,[83] and Eudemus made great advancements in anatomy—particularly respecting blood. First on the list is Herophilus, who advanced past his teacher, Praxagoras, in giving a *structural* difference between arteries and veins. Besides their point of origin, Herophilus showed that the arteries have more layers in their tubular construction than do veins:

> In all the other parts of the parts [of the body, as opposed to the artery that goes to the lungs], when an artery is equal [in size] to a vein, the thickness of their coats is not equal. Rather, it is different to such a degree that Herophilus seems to have conjectured correctly when he stated that the artery is six times as thick as the vein (tr. von Staden).[84]

There is the question of *in general* and *by nature* in the attribution of thickness. It seems that what Herophilus is asserting here is that, in general, the walls of the arteries are much thicker than those of the veins. He noted that his teacher, Praxagoras, said that vessels emerging from the right ventricle of the heart were veins and from the left ventricle were arteries. This led Herophilus to identify the pulmonary artery as the artery-like vein (because it attaches to the right side) and the pulmonary vein as the vein-like artery (because it attaches to the left side).[85] This allows Herophilus to maintain the bilateral symmetry of Praxagoras (a *phusis*$_2$ notion)[86] along with the more complicated material anatomy that he had discovered.

The arteries themselves carried blood and *pneuma*.[87] This represents a departure from Herophilus' teacher who asserted the arteries were pneumatic tubes that turned into nerves. However, on the subject of nerves, Herophilus seems again to have departed from his teacher in asserting a separate system of nerves that began in the brain rather than the heart.[88] Herophilus did careful dissections of the brain, separating the ventricles with more precision. Some of this is evident in his penchant for naming, for example, the pineal gland,[89] and giving more specification to the *rete mirabile*.[90] His work on the nerves included his distinguishing motor and sensory nerves[91] and giving much more specification on cranial nerves.[92]

Herophilus also differs from his teacher concerning the pulse. He did not accept the sole independent power (*dunamis*) of the arteries that Praxagoras had put forth. Instead, he set the power of pulsation in the heart.[93] Harris argues that the three terms that were used by Praxagoras for "pulse" became more specialized, with "palpitation" (*palmos*) used to cover all irregular motions of the arteries in a general sense—including unnatural distention. "Tremor and spasm" (*tromos* and *spasmos*) were removed from the vascular vocabulary. What was needed was regularity, and Herophilus used *sphugmos*. In its primary usage, it was restricted to the heartbeat (a fortiori the modern sense of pulse). This came to include the effects of the heartbeat: the sending out of the *pneuma* that traveled through the blood in the arteries, which caused periodic expansion (contraction follows after the force of the expansion has passed). This was not a pump, but a periodic release that is linked to the periodic expansion of the arteries. He considered it to be regular like a clock and thus it could be used as a role in diagnosis[94] (a role that it kept until this very day).

We have less reliable evidence regarding Erasistratus. Unlike Herophilus and like Praxagoras, Erasistratus believed that the arteries contained no blood, only *pneuma*.[95] Being a practitioner who was keen on giving a full, mechanically oriented, material explanation, he felt that he had to account for the fact that when someone is cut in an artery, blood issues forth instead of *pneuma*. His answer to this relates to a theory of vacuum (*kenon*), which originates in the heart. It should be noted here that Erasistratus was a keen observer of the heart and may have been the first to identify all the valves of the heart. Their function was to keep the heart rhythmic like the bellows of a blacksmith.[96] This image of a bellows bestowing *pneuma* also allows for the possibility of a vacuum. For anyone who has ever tried a bellows, the first stroke is to pull apart the levers (diastole).[97] This creates a vacuum. Then one exerts force to push out the air by contracting the bellows' levers (systole). This later stage sends out the *pneuma* into the body. This sounds a bit like characterizing the heart as a pump, which was the major hurdle that Harvey had to clear as he came up with his theory.[98] This may be a bit of a stretch because there was also the principle of *attraction* working in which the *pneuma* and the blood operating within the body were *pulled out* of the heart into the body.

The two separate systems (blood and *pneuma*) generally did not interrelate except at the level of the capillaries. However, they ran close together so that a cut of the artery could create a strong vacuum and *pull* blood from a contiguous vein into the chasm. This would give the illusion of blood flowing out from the artery.[99] There are various objections that can be raised to this account—particularly that clean cutting of an artery yields pulsating blood evacuation. This and other issues make Erasistratus' account less convincing.[100]

However, following in the tradition of Hippocratic and Aristotelian critical Empiricism, Erasistratus was inclined to a sort of experimentation.[101] I will call this sort of experimentation *the single perturbation account*. In the single perturbation account, one alters nature in a single instance and then sits back to interpret what has happened. The drawback of this approach is that it may exaggerate the significance of later event episodes via *post hoc ergo propter hoc*. It is unclear just how the single perturbation affects the complete dynamics of the event.[102] In this instance, Erasistratus was exploring the operation of the pulse. One key question is whether the pulse was caused by a material property of the walls of the artery or caused by the *pneuma* that was bellowing out from the heart. The result was a single perturbation account experiment. Galen describes it this way:

> You must expose one of the big arteries near the skin, like the one in the groin, which is the one I use most frequently to perform this experiment. You must tie a ligature round it in its upper part. Then with the fingers of your left hand compress it as far as possible below the ligature, but above the point where it throws off any large branch. In the portion between these two points, you make a straight longitudinal cut, large enough to be able to insert a hollow tube between the ligature and the fingers of your left hand. . . . It is clear that there will be no hemorrhage from the artery that is being divided, since the higher portion from which the blood flows has been shut off by a ligature, and the part below the cut no longer pulsates . . . [insert the tube and remove the ligature] you will observe that above [the tube] the artery continues to pulsate but below [the tube] there is no pulse (tr. based upon Harris).[103]

This procedure, Erasistratus contends, shows that the insertion of the tube removes *pneuma* by giving it an avenue of escape. Thus, the question is answered! If the pulse were a property of the structure of the arteries themselves, then there would be no difference in the pulse from above or below the ligature/tube insertion. Because of both the ligature and the cannula, the *pneuma* from the heart would be regionally emptied. This *emptying* stopped the source of the pulse.

However, Galen demurs. Of course, Galen did not hold that the arteries only carried *pneuma*; therefore, the whole dynamic is rather different.

> Erasistratus' statements on the movement of the arteries [pulse] is entirely false . . . before the cessation of movement in the denudated artery in those sections below the ligature . . . it is possible to see the whole artery moving at the same time and not one part of it first and the other afterwards as Erasistratus alleges . . . the fact that after the ligature has been tied pulsation ceases in the parts of the artery distal to the tube shows clearly that the power of dilation of the arteries is communicated to their coats by the heart and not by the matter flowing out through the tube (tr. based upon Harris).[104]

Galen believes that Erasistratus is committed to a position: the heart as the origin of the pulse via a pushing out of *pneuma*, which would make the pulse into a rolling bubble going forward. But Galen wants the arteries to pulse all at once, meaning that it comes from the various coverings of the arteries. These conclusions are rather different, and the experiment lends itself to supporting *each position* according to the side conditions one imposes. What makes this procedure so interesting is that it constitutes a very controversial aspect of the single perturbation account style of experimentation. The one and the same action (which was repeated often by the likes of Vesalius[105] and Harvey[106] with some small variations) was interpreted differently by each author. Some of this had to do with questions on cutting and the ligature (what we now call the laboratory technique). But, in general, it is more than this, and as such goes to the very problem of naïve enumerative induction. The word "naïve" is used both because only the naked senses are employed and because it lacks the structure of the hypothetical-deductive method with its attention to dependent and independent variables. The "enumerative" part has to do with repeating the procedure whereby the general principle is supposed to leap out. Aristotle's *Posterior Analytics* II.19 is the emblematic case of this. However, modern scientific theory uses principles of inductive logic that more mechanically reduce the level of subjective judgment by reference to laws of mathematical probability.[107]

In the end, we have in Erasistratus a biomedical practitioner who sought to ground his theory in experimentation along the line of the single perturbation account. This creates a link to those who, wary of theory, appreciate the possibility of anatomical and physiological (via the single perturbation strategy) knowledge. But as we have seen earlier, this line of inquiry presents the possibility of various interpretations: enter the Empirics and Methodists.

SKEPTICISM

Before we turn to the Empiricists and Methodists, we must first examine the underlying attitude that affects Empiricism (proximately) and Methodism (remotely). This attitude is skepticism. 'Skepticism' is a difficult concept. It has many different sorts of meanings, both in its classical Western context and in how that context has been interpreted for modern times. One way to get into this controversy is to step into the shoes of a modern and then backtrack. According to David Hume (an authoritative source on how modern philosophy views Skepticism) there are two general categories of Skepticism: (1) $skepticism_1$ refers to radical doubt (meant as obliteration of all that had been assumed) and (2) $skepticism_2$ concerns mitigated doubt (meant as a springboard to future knowledge).[108] The *radical doubt* tract follows Descartes' thought experiment in the *Meditations on First Philosophy*. Under this strategy, one engages in doubting everything possible. Of course, for Descartes, one cannot doubt the grammatical structure of language that allows all doubt to go forth so that all transitive verbs require subjects and objects.[109] And this caveat has caused many to view Descartes' epistemological exercise as circular.[110] Much of the modern understanding of *skepticism* began with the Cartesian possibility of radical doubt.[111]

However, Hume discards radical doubt in favor of *mitigated skepticism*. Under this account, the philosopher ought to employ epistemological standards of due diligence when confronting both novel events and accepted paradigms of explanation. This follows a model of epistemology that seeks to make as few errors as possible when trying to make sense of the world. I have characterized such an approach as the *careful scrutiny approach*.[112] Under the careful scrutiny maxim, one seeks to minimize epistemological error at all cost. The upside of this method is that (in theory) it minimizes mistakes. Physicians who adopt this approach will be inclined to stick to what they were taught and handbooks of what has worked for others. It is a very conservative epistemological disposition.

A third approach to skepticism captures more of what in the ancient Greek world was meant by *skepsis*. This word definitely has an empirical slant, but it also has a slant that is reticent of closure.[113] This sort of skepticism ($skepticism_3$) requires continual consideration that bristles against epistemological closure. It is an extreme case of the careful scrutiny maxim. Because the careful scrutiny maxim is so directed toward making *no* epistemological mistakes (or as few as possible), there is a bias about not making firm, permanent commitments toward proposition x being necessarily true. By keeping everything open, the skeptic seeks a tranquility, *ataraxia*.

An example of $skepticism_1$ in the ancient world would be Parmenides' monism and Zeno of Elea's defense of the same through his paradoxes of motion.[114] These arguments try to set out that the empirical world of change and motion that we observe is rationally incoherent and so must be rejected in favor of a world we do not see but rationally posit as coherent. In some ways, this is methodologically similar to Descartes' approach. It is not surprising that the reactions to Zeno were similar to what Descartes faced. Heraclitus also falls into this camp with his assertion of the unseen *logos* as being true and sense impressions being set out in a context of opposites via the metaphor of an ever-moving stream.[115] These are the most extreme forms of skepticism, and they align themselves with an epistemological certainty that connects with D-□ (see Chapter 2, this volume).[116] Because of the chasm that is created by the initial doubting-as-obliteration, the only way to closure is what I call the selected faith maxim.[117] Under this strategy, the scientist wants to maximize his garnering of all possible truths—even if it means letting in some minor errors along the way. This sort of epistemological disposition seeks to discover all that can be known. The physician of this sort is always seeking to stretch what is known about the human body: in health and in disease. Such an individual can be called a progressive innovator.

For skepticism$_2$, where critical questioning is really a springboard toward finding underlying truths (for the most part, D-◊), I would put forward the various dogmatic philosophers—particularly those inclined towards Aristotelian critical Empiricism. These biomedical practitioners use gaps in the explanatory framework to launch critical inquiries based upon Aristotelian explanatory principles (as per *Apo.* 11.19; *PA* I; and various *nomos* pronouncements such as "Nature does nothing in vain," bilateral symmetry, et al.). The Dogmatist practitioners set out earlier fit into this category. They base their D-◊ epistemological pronouncements upon observation and experience that are structured by a self-consciously adopted methodology. Unlike the selected faith maxim method of skepticism$_1$, they follow the different epistemological tract: *the careful scrutiny maxim*. Under this maxim, the scientist seeks to minimize error at all costs. It is my conjecture that modern science follows this epistemological strategy.

In skepticism$_3$ we have various practitioners about whom we have general testimony—such as Carneades—but few texts to support critical examination.[118] However, we do have texts from Sextus Empiricus, who was also a physician. He said in the *Outlines of Pyrrhonism*:

> Skepticism is an ability (*dunamis*), or mental attitude (*nooumenon*), which opposes appearances to judgments in any way whatsoever, with the result that, owing to the equipollence of the objects and reasons thus opposed, we are brought first to a state of mental suspense and next to a state of "un-perturbedness" or quietude (*ataraxia*). [119]
>
> The main basic principle of the skeptic system is that of opposing to every proposition an equal proposition; for we believe that as a consequence of this we end by ceasing to dogmatize.[120]
>
> When we say that the sceptic refrains from dogmatizing (*dogmatizein*) we do not use the term "dogma (*tou dogmatos*)," as some do, in the broader sense of "approval (*eudokein*) of a thing" (for the skeptic gives assent to the feelings which are the necessary results of sense-impressions and he would not, for example, say when feeling hot or cold "I believe that I am not hot or cold") . . . Moreover, even in the act of enunciating the Sceptic formulae concerning things non-evident—such as the formula "No more (one thing than another),"

> or the formula "I determine nothing," or any of the others . . . he does not dogmatize.[121]

These roughly proximate passages set out a theory of skepticism$_3$ that may be a very justifiable response to a science that is so nascent. Several points are clear: (1) Sense impressions are accepted as accurate (one can say with assurance whether he is hot or cold); (2) Theories that might be set forth about the underlying causes of these sense impressions are no more true than those that are inconsistent with such theories (you feel hot because of too much blood or black bile as opposed to you feel hot because of too little blood or black bile); and (3) No self-referential pronouncements may be made against such a position (one is not permitted to overlay a self-referential paradox).[122] Thus, the skeptic$_3$ can admit empirical symptoms of the patient and can entertain responses, but may not judge these responses as the *ultimate*, proper response.

It is important when viewing the skeptic$_3$ claims that we distinguish between epistemology and ontology (a branch of metaphysics). Epistemology seeks to enunciate rules about the sources of knowledge (such as pure rational inquiry, critical Empiricism, or naïve Empiricism), the methods used to approach those sources (the careful scrutiny maxim or the selected faith maxim), and what we think we can know as a result (the objects of knowledge: ontology). In this last category, the line between the mode of knowing (epistemology) and what is known (ontology) is very thin. For our purposes, let us generalize on the object of knowledge as being *phusis* in its three forms (see Chapter 1, this volume). The way that one envisions a reliable way to know will make clear the proper way to approach the object of investigation. In the case of *skepticism$_1$* one is making radical claims about doubting empirical sources that can only be resolved when the object concerned is nonempirical to begin with—the gods and supernatural objects, *phusis$_2$*. In the case of *skepticism$_2$*, which is meant as a springboard to reasonable general principles that can be used to provide a rather broad *explanans* to a rather diverse *explanandum*, we are presented with *phusis$_1$*. This will be the approach of the Dogmatists. Finally, in *skepticism$_3$* we have reluctance at embracing *any* single theoretical *explanans* because either the *explanandum* is too complex (a variant of the "three-body problem" in Newtonian physics)[123] or because there is an inherent limitation in human reason's ability to create a complete *explanans* (an epistemological limitation on what can be known).[124] Some of these relations can be set out in the following table.[125]

Table 4.1 represents how a number of distinctions set out in Chapters 1 and 2 in this volume relate to the modes of skepticism set out here. Together, they create a general profile that can be useful to understand different sorts of practitioners of biomedicine between the earliest natural philosophers going forward until Galen (roughly the scope of this book). Briefly, let's review this composite.

Table 4.1 How Forms of Skepticism Affect the Profile of the Biomedical Practitioner

Skeptic-Type	Epistemology	Ontology	Type/Token	Modality/Practice
$Skepticism_1$	Selected Faith Maxim/ Explanatory-type β	$\textit{Phusis}_2$	Type-oriented	D-□/Philosophical
$Skepticism_2$	Careful Scrutiny Maxim/ Explanatory-type α	$\textit{Phusis}_1$	Type and token equally	D-◊/Philosophical
$Skepticism_3$	Extreme Careful Scrutiny Maxim/ Explanatory-type β	$\textit{Phusis}_3$	Token-oriented	E-◊/Clinical

$Skepticism_1$ falls in line with radical doubt (generally concerning the reliability of sense impressions about the world). Many of the earliest practitioners of biomedicine would fall into this camp (see Chapter 1, this volume) along with Parmenidean monists and Heraclitean advocates of *logos* over sense impression. Since empirical knowledge is held in doubt, the principal source of knowledge would be the selected faith maxim concerning possible emergent properties (explanatory-type β—see Chapter 2, this volume) in a $\textit{phusis}_2$-dominated conception of nature. Since $\textit{phusis}_2$ is highly connected to religion and its tangible, relatively unpredictable application to problems in everyday life (magic), the epistemological emphasis is type oriented (whether this is a particular god or one of its rationally asserted properties or forms). Because the object of examination is a type and because the selected faith maxim was used, the modality of the outcome will be philosophical certainty. There is not much room for discussion here, since all contrary instances can be explained away via the rather loose explanatory type-β mode.

$Skepticism_2$ is the domain of the Dogmatists. Many of the Hippocratic Writers, Aristotle, Diocles, Praxagoras, Herophilus, and Erasistratus would fall here. They would employ the careful scrutiny maxim along with explanatory-type α that seeks to create a mechanical process using material agents to produce a result that, in principle, can be described without reference to inscrutable agents. It makes reference to types and tokens equally since it is a theory of *critical* Empiricism. The end result is some sort of structured inductive process that starts with experience and produces law-like outcomes (subject to *epi to polu*, for the most part). These outcomes are not certain because of *epi to polu*. Thus, the end result is a D-◊ set of explanatory principles that fit together systematically. The midpoint of D-◊ can be the result of very complex and sometimes changing phenomena. This is the model that contemporary science has generally adopted.

Skeptic$_3$ describes the general background of the Empiricists. Under this approach, the practitioner adopts a rather extreme form of the careful scrutiny maxim. This added feature would include an externalist realism that asks searching questions about how naïve sense data might be misinterpreted as generally applicable (thus creating a distrust for inductively based principles that seek wide applicability).[126] This leads to giving special credence to the individuality of the patient and his or her particular circumstances (in the case of the Empiricists) and of particular forms of diseases as presented. The patient for the Empiricists is *almost* sui generis (the clinical approach). This means that some patterns may be useful in examining causes of the disease itself via symptoms, but clinical individuality (the token perspective) leads to a variable state of affairs (E-$\Diamond$) that requires the on-the-scene judgment of the attending physician in order to act with clinical discretion.

It is the contention of this chapter that this variable understanding of Skepticism will shed considerable light on what the various approaches to biomedicine are up to. We have dealt with the Dogmatists. Now we will briefly examine the Empiricists and the Methodists.

THE EMPIRICISTS

What were the Empiricists (*empeirkoi*) about? Well, to begin, it has to do with the root term, *empeiria*, which means experience, but what does "experience" mean? Does it refer to uncritical interaction with the world or something more robust along the lines of Aristotle's critical Empiricism? This is a very big question.[127] The answer goes back to Celsus quoted earlier: the evident and the hidden. The Dogmatists were keen on discovering the hidden, underlying mechanisms that had broad explanatory power. But the Empiricists were reticent because they were advocates of skepticism$_{3.}$ Their response to the Dogmatist theories was, "Not so fast."

I follow those who situate the Empiricists in the Hellenistic Medicine period (largely as a reaction to the Dogmatist anatomists Herophilus and Erasistratus) in the person of one of Herophilus' students, Philinus of Cos around 260 BC.[128] The Empiricists did not believe in the hidden causes because of epistemological reservations (see the discussion on skepticism$_3$ earlier).[129] Instead, they relied upon *peira* and *empeira* (the former being more connected to trial and error). In this way, von Staden likens them to B. F. Skinner's behaviorism, which used operant conditioning to create trial-and-error, physically evident results that belied "hidden" psychological motivations of the so-called folk-psychologists (those in the Freudian tradition).[130] Since there were so many possible epistemological objections to the hidden via *skepticism$_3$*, the alternative, viz., *the evident* was chosen. This was done for clinical reasons of efficiency. After Herophilus and Erasistratus, this approach became dominant (partially also due to *phusis*$_2$ advocates and those who had religious and moral objections to dissection).[131]

If one were to create a general approach for the Empiricists, it might be the so-called *tripod*: (1) keen observation of the patient by the physician, *autopsia*; (2) some lore of what had worked in the past (either in the practitioner's firsthand experience or via secondhand experience—the writings of some other practitioner), *historia*; and (3) a skepticism$_3$ approach toward the attitude of similarity of connection.[132] These deserve a little more attention. First is the keen observation of the patient (firsthand observation). This includes listening to the reports by the patient of his own condition along with the physician's own observation. However, this is also a bit controversial concerning blood. This is because the pulse is the one of the primary "evident" inputs that is available to the physician. However, it is impossible to totally expunge theory from observation (even if the theory is not properly recognized as such). Herophilus had a rather complicated account of the pulse that distinguishes the beginning of the heart's systole from its quiescence (which varies greatly from newborn babies to old men). In counter distinction, the Empiricists do not consider the diastole because it is not observable with the fingers. Instead, they simplify by measuring only beat and interval.[133] This a conscious choice they make to eschew the more complicated theory of pulse set out by Herophilus that arose from his visually based understanding of the heart's activity. But if you reject dissection in either of its forms, then you will likewise reject this more nuanced account of the pulse as a diagnostic tool in favor of a simpler model. This is because those who view the body as an entity that, in principle, can be understood mechanically will be drawn to anatomy and physiology. Those who demur do so because they view this asserted precision as illusory: better to concentrate upon the patient before him and the macro symptoms. Interior diagnostic tools (such as the pulse) lead the physician, by necessity, to theories of the hidden—Dogmatism.

The second leg of the tripod refers to the past. First in line would be the attending physician's own personal firsthand experience and his knowledge of what worked and what didn't work in *similar* circumstances. This evidence-based medical approach[134] was highly reliant upon drug treatment (which, along with lifestyle choices and regimen, constituted the general purview of the physician). Since regimen is largely a preventative medicine technique, it was not of value in situations in which the patient might be in crisis. The only other technique, manual manipulation (including cutting—sometimes called surgery), is largely constrained for the Empiricists to versions of massage since invasive cutting inclines one to the acceptance of humor theory.[135]

However, when the attending physician is either new at the job or encounters a case that is novel to him, then there is recourse to secondhand trial-and-error accounts. These might be the more clinically oriented accounts from earlier writers[136] or via oral transmission (much as folk-medicine accounts were later transmitted in medieval Europe generally via the women in the family).[137]

The final leg of the tripod refers to *similarity*. This is a very tortured philosophical concept. It refers to a correspondence theory of truth. In the correspondence theory, some instance, x, is true just in case x resembles some model that is assumed to be a proper instance.[138] There are two basic flavors of this theory: (1) the one-many interpretation of Plato's Theory of Forms where when some token x is compared to the Form of X—the more similar it is, the more true it is; and (2) natural typology where when some token x is compared to an accepted natural type of x, then x is more true as it approaches verisimilitude to the accepted type. In the first instance, we must posit a separate realm of nature along the lines of *phusis*$_2$. This, of course, was not the direction of the Empiricists because it included the inscrutable. In the second instance, one is comparing some instance of a fever (a token) with all the *tokens* of fever one has experienced (directly or secondhand). In the judgment of the attending physician, the token is tied by common properties to some other token instance. This act of identification constitutes the similarity leg of the tripod.

However, the act of identification and the epistemological barriers of *skepticism*$_3$ also are the grounds for discarding the work of the anatomists Herophilus and Erasistratus. For starters, a live body is not the same as a dead body. For all one knows, these major dissimilarities might lead to major error. Even vivisection creates a very artificial context that might lead to no useful information. Thus, if there are moral and religious objections to dissection and vivisection, and if there is a very real possibility that the end product is *inaccurate*, then these factors incline the Empiricist practitioners to avoid this data. Roughly speaking, this application of the careful scrutiny maxim allies itself to the "do no harm" tenet in medicine (cf. the Hippocratic Oath).[139]

But implicit within the notion of *similarity* is a notion of connection. A token "x" is similar to a token "y" on the basis of some connection, F (where "F" is a property that they both share). Galen, in his work *An Outline of Skepticism*, suggests that the Empiricists put forth a five-part continuum by which x may be similar to y: (1) when the shared trait always occurs, (2) when the shared trait occurs for the most part, (3) when the shared trait occurs half the time, (4) when the shared trait occurs rarely, and (5) when the shared trait never occurs.[140] This mix of qualitative and rough-quantitative criteria provides some substance to what is meant by *similarity*.

However, because the Empiricists do not accept type-oriented Dogmatic pronouncements on hidden truths (scientific laws), even in the first category in which the connection occurs all the time, there is no necessity to this connection.[141] The *sign* (*sema*) can have power of self-validation (epistemologically).[142] This validation of the sign is either natural (based upon the first form of correspondence theory of truth) or conventional (based upon the second form of correspondence theory of truth).[143] It is clear that the Empiricists intend the latter, since the former would force them to be Dogmatists.

However, Galen (a Dogmatist) argues in *On Medical Experience* that every patient presents in a unique way. Given this, how can the Empiricists claim to find *actual* properties that x and y share? Often, two patients will share many properties, but most of these will be accidental and not medically relevant. How can the theory-eschewing Empiricists authentically assert which similarities are relevant and which are not—without Dogmatic theory?[144]

Despite this rebuke, Galen did recognize Heraclides of Tarentum (an Empiricist) as a learned practitioner.[145] However, the great reliance upon drugs for treatment in a trial-and-error fashion also inclined many Empiricists toward $phusis_2$ in the use of charms.[146]

The reason that the Empiricists were not so keen to delve more into blood is because blood is part of the hidden and so really out of bounds. Only the pulse (in a simplified account) is given to the clinical physician practitioner. However, there is the treatment of bloodletting that was practiced by some Empiricists (see Chapter 5, this volume).

THE METHODISTS

In the clinical setting, what is the proximate center of attention? There is, of course, a patient who has come to you for help. But how do you create an epistemological framework to deal with your patient? One way is to focus on the disease. Of course, this was also the strategy of the Dogmatists. They focused upon the disease in a manner that emphasized the *causes* of disease with reference to characterization of $phusis_1$. However, another approach is to streamline the process—bypassing anatomy and speculation upon the material structure of human beings and how they might get out of sync. In order to avoid the philosophical debates of the Dogmatists (which skeptically oriented physicians found tedious and unhelpful), and to avoid the Empiricist complicated, clinically oriented therapeutic metrics, the Methodists offered a simpler handbook approach based upon the commonalities of diseases.[147] Disease is presented as constricted, unconstricted, or mixed.[148] Diseases themselves were either chronic or acute (with the former being too constricted and the latter too unconstricted).[149] This led to treatment strategies that accepted the *diatritus*, ongoing three-day cycles, in order to track the path to recovery.[150] This streamlined approach could be franchised on a large scale rather efficiently, which is why learning the therapeutic approach, called "the method" (*methodos*), was a path that could be completed in as little as six months![151] This was the quickest apprentice model for physicians among the various schools. It was no wonder that such efficiency was greatly appreciated in the Roman Empire that valued the combination of simplicity and results.

The emphasis upon results over theory is also a characteristic of one form of scientific pragmatism. Pragmatism is outcome oriented. Whether some posit is true or not is dependent upon whether using that posit in a particular context gets you what you want: it works. So if we use the constricted/unconstricted/mixed model and get results, then that is truth. This goes against the correspondence theory of truth model of the Dogmatists.[152] The long-standing argument against scientific pragmatism is that something may "work," not for the reasons given, but for other reasons unknown to the practitioner—this is called a *sumbebekotos*, accidental, inference by Aristotle (*Meta*. I, 4, 8). However, such distinctions were of no practical interest to the Methodists.[153]

Since there are so few existent complete texts by Methodists (which is due to Galen's dominance),[154] we must turn to Soranus to see how this theory plays out in explanations about two cases concerning blood. In the *Gynecology* Soranus addresses the female discharging too much blood due to either a difficult labor, miscarriage (*ektrosis*), or various ulcerations or irregular, heavy menstrual flow.

In both cases we have a discharge of too much blood. In order to get some background on the discharge of blood in women, we must briefly return to *ichor*. As the reader may remember, in Chapter 1 in this volume, *ichor* was set out as a connection to the blood of the gods. It provided support for $phusis_2$ explanations. In Chapter 2 in this volume, there was more nuance as some common usages of *ichor* as a thin fluid associated with animal meat juice stood side by side with other usages by the tragedians as still being blood of the gods.[155] In the writings of Soranus, the bestial (lower-than-human) meaning is primary and is used to explain disease:[156]

> Re: Menstrual blood v. *ichor* (Soranus I. 4. 19)
> Menstrual blood is depicted as pure while *ichor* is corrupted [like] animal meat. When the menstrual flow comes from the uterus, it is pure blood. When it comes from the vagina, it is tainted and displays characteristics of *ichor*.

> Re: Fertility and the proper moisture of the uterus (Soranus I. 9. 34)
> Blood is set out as providing the proper moisture while *ichor* is an ersatz moisture that will not prepare the uterus properly.

> Re: Spontaneous abortion (Soranus I. 18. 59)
> "a watery discharge appears . . . an *ichorous* or sanguineous fluid—like the water in which meat has been washed." When this appears in a pregnant woman, a miscarriage is about to occur. Once again *ichor* is a sign of *para phusin*.

> Re: Skin exanthemata and itching (Soranus II. 25 [45]. 52 [121])
> "When the *rheumatismos* [flow of bodily fluids] and the great discharge of *ichors* have ceased . . . we resort to olive oil . . . and with the white of an egg beaten combined with a moist wax salve [an astringent]."

> Re: Uterine Prolapse (Soranus IV [XX]. 36 [85] "Euryphon makes the patient hang by her feet from a ladder for a whole day. . . . Euenor inserts ox meat into the vagina not knowing that the *ichors* produced by the putrefaction will, by their pungency, cause ulcerations."

In each of these cases we see *ichor* to be an agent of the unconstricted. The remedy is to balance the unconstricted with various constricting agents such as a vinegar and papyrus potion that is rubbed on the labia and the orifice of the vagina (vinegar is an astringent, and papyrus is part of the delivery mechanism meant to keep the mixture in place).

In severe cases, a vaginal suppository is used consisting of ashes of a sea sponge soaked in raw pitch.[157] Ashes are astringent, and pitch is an effective device to hold the suppository in place.[158] These treatments are meant to counteract the unconstricting agents (such as *ichor*), which are *causing* the miscarriage, ulcerations, or heavy menstrual flow. (We have to be careful about characterizing these agents as *causing*. This is because "causation" is a by-product of the Dogmatists and their skepticism$_2$ approach [see Table 4.1]). But the Methodists *are* asserting three *type-relations*: the constricted, the unconstricted, and the mixed. Type-token relations are explanatory, so they function causally—whether the Methodists admit it or not. The Methodists want to say that their categories are merely self-evident commonalities that are revealed during the three-day cycles of disease. But this is incoherent, as Galen and Celsus point out.[159] It ignores the distinction between what is physically evident in the body (Empiricism and Dogmatism) and what might be hidden (Dogmatism). It also brings in a notion of types by the backdoor (Dogmatism). For this reason, it seems useful to categorize the Methodists as in between the Empiricists and the Dogmatists. Using the categories from Table 4.1, the Methodist would be *skepticism*$_{2.5}$, between explanatory types α and β; a semi-careful scrutiny maxim advocate, *phusis*$_3$, they use (but disavow) the type and acknowledge the token (but instead of this being the patient, it is the disease): these all result in classifying the Methodists as between D-◊ philosophical and E◊ clinical.

All three of these approaches to medicine—Dogmatists, Empiricists, and Methodists (seen in the context of the various forms of Skepticism)—had their advocates in the ancient world. However, it was Galen who put the most significant stamp on ancient biomedicine as he considered various approaches that came before and forged his own amalgam.

NOTES

1. One must be very careful when discussing "programs." As Vivian Nutton has mentioned to me, Diocles may not have had many *followers*. This does not diminish his achievement in the history of ideas (to which this book is committed), but it is important to note this within the historical context.

2. See Chapter 2 in this volume on the various grounds of modality. For a further discussion of the strong influence of Aristotle on later biomedical writers, see James Longrigg, "Medicine and the Lyceum" in Philip van der Eijk, H. F. J. Horstmanshoff, and P. H. Schrijvers, eds. *Ancient Medicine in Its Socio-Cultural Context: Papers Read at the Congress Held at Leiden University, 13–15 April, 1992*, vol. 2 (Amsterdam: Rodopi, 1995): 431–446.
3. Galen, *On Medical Experience* 13, 4, tr. Walzer and Frede 1985: 70. Etymologically, δόγμα refers to opinion or belief (Plato, *R.* 538c), but more especially to philosophical doctrine (Epicurus, *Nat.* I, 4–7; I, 5–28). Since this latter sense gives the notion of not only 'belief' but 'true-belief,' I will refer to these authors as Dogmatists.
4. Hui-Hua Chang, "Rationalizing Medicine and the Social Ambitions of Physicians in Classical Greece" *Journal of the History of Medicine and Allied Sciences*. 63.2 (2008): 217–244 believes there is another factor at work, viz., the low social standing of physicians. Because of this, physicians adopted speculative, rationally based theories because these were attractive to the wealthy (who also patronized the sophists). It also created a medical practice that emphasized regimen (preventative medical care aimed at maintaining health)—and thus offered more social contact between the physician and his wealthy clients.
5. Frede (1985): xxxii–xxv.
6. One group that this chapter does not discuss is the so-called Pneumatic School, who explored further the way that *pneuma* came into and interacted within the blood—for a discussion, see Harris (1973) 235–251.
7. Steckerl (1958), fragment 13, tr. Steckerl. Fragment from Galen, *de uteri diss.* 10, II p. 905 Kühn. Nutton (2013): 122–125 also puts Diocles after Aristotle and as a contemporary to Praxagoras, cf. Van der Eijk (2000–2001) II xxxi–xxxviii.
8. Anonymous of Brussels 1–8 (p. 208, I-213, 14 Wellman), cf. fr. 40 Van der Eijk (2000–2001). It should also be noted that Diocles was a very comprehensive physician. Philip J. Van der Eijk in "The Systematic Status of Therapy in the Hippocratic Corpus and in the Work of Diocles of Carystus" *Aspetti della terapia nel Corpus Hippocraticum : Atti del IXe Colloque International Hippocratique* (Pisa, 25–29 September, 1996) in *Medical History* 45.1 (2001): 389–404, argues that of the three principal areas in ancient medicine—regimen (dietetics), medicaments (pharmaceutics), and manual intervention (surgery), Diocles was widely considered to be a master of all these areas, making him truly the second Hippocrates.
9. Nutton (2013): 120–123.
10. Galen, 2, 281–282; Van der Eijk (2000–2001) collection of the fragments sets this as 17 cf. fr. 24.
11. Nutton (2013): 122.
12. Philip Van der Eijk (2000 and 2001) has become the standard source replacing Wellmann (1901). However, I will cite both sources according to individual circumstances.
13. Anonymous Bruxellensis, 44. (cf. Thomas Bricot).
14. Harris, 104, cf. Nutton (2013):122.
15. See Chapter 1, this volume, on the various versions of *phusis*.
16. Wellmann, fragment, p. 14.
17. Harris, 105.
18. I am following Van der Eijk (2000) [Diocles fr. 177] and Nutton (2013): 143 in putting Diocles after Aristotle contra Wellmann (1901).
19. Diocles fr. 27, 51, van der Eijk.

20. See Chapter 2, this volume.
21. Nutton, (2013): 124.
22. Van der Eijk suggests that this may be the result of some *pneuma* being "psychic" *pneuma*: Diocles, fragment 74, 80, cf. 101, 107.
23. Galen, *De usu resp*. I, K. IV: 471.
24. Plato, *Timeas* 79 b & c.
25. Vindicianus, Wellmann, *Fragment*, p. 228, cap. 32.
26. Diocles fragments 180–181, Van der Eijk, cf. Pliny XX, 19, 34, 52, 106, 139, 219. Nutton suggests that this emphasis came from Diocles' father, who wrote a work on this subject.
27. Cf. the fragments of the Anonymous Parisian of Fuchs.
28. Among these also include Mnesitheus and Dieuches (from Athens), Praxagoras, Phylotimus, and Pleistonicus (from the island of Cos), Galen, *On Examining the Physician* 5, 2: *CMG* Supplement Or. 4, 69 *On Medical Experience* 13, 4. See Vivian Nutton, "Galen Avoiding Distress" translation in P. N. Singer, ed. (forthcoming) *Galen: Psychological Writings* (Cambridge: Cambridge University Press). A slightly more nuanced classification has been offered by Ravi Sharma in "Diocles of Carystus on Scientific Explanation" *The Classical Quarterly* 62.2 (2012): 582–602. Sharma examines Diocles fragment 176 (Van der Eijk) in great detail and concludes that Diocles exhibits shades of Medical Empiricism in this text.
29. Praxagoras fragment 45, Steckerl (1958).
30. Praxagoras fragments 18, 21, 27, 32, 86, Steckerl (1958).
31. Celsus, Prooemium, 8, cf. Kurt Bardong, Beiträge apply Hippokrates-und Galenforschung. (Göttingen: Vandenhoeck & Ruprecht, 1942)
32. Praxagoras fragment 1, 98, Steckerl (1958).
33. Anonymus Londinensis, *Anonymi Londinensis ex Aristotelis Iatricis Menoniis et aliis medicas eclogue*. Hermann Diels, ed. (Berlin: Suppl. Aristotelicum, 1893): 8, 10, 35. However, Diethard Nickel in "Hippokratisches bei Praxagoras von Kos?" in Philip J. van der Eijk, ed. *Hippocrates in Context: Papers Read at the XI International Hippocrates Colloquium, University of Newcastle on Thyme, 27–31 August 2002* (Leiden: Brill, 2002): 315–323—that Praxagoras was really rather original and not merely modifying Hippocratic doctrine.
34. Praxagoras fragment 70, 71, Steckerl (1958), Fuchs: 541, 549.
35. Praxagoras fragment 27, 45, Steckerl (1958).
36. Praxagoras fragment 24, 37, 45, Steckerl (1958).
37. Plistonicus fragment 1, Praxagoras fragment 16, Steckerl (1958).
38. Praxagoras fragment 88, Xenophon fragment 3, Steckerl (1958).
39. There are at least two physicians who are named "Praxagoras." I am following Steckerl here to use his compilation to the Coan school. A note of caution on this score comes from Nutton, who has pointed out to me that school traditions are often late and faulty.
40. Praxagoras fragment 22, Steckerl (1958).
41. Praxagoras fragment 22, Steckerl (1958).
42. Plistonicus fragments 1, 4, Steckerl (1958).
43. This attribution may merely be a logical one since it is probable that Dogmatists were not identified as such at this time.
44. Praxagoras fragment 15, Phylotimus fragment 1–2, Steckerl (1958).
45. R. Fuchs, "Anecdota Medica Graeca" *Rheinisches Museum* 49 (1894): fragments 3, 4. More complete work on the anatomical distinction between arteries and veins was made by Praxagoras' pupil Herophilus.
46. As it works out, all vessels that connect to the heart from the right ventricle were veins (though they originated in the liver) and all vessels that connect to the heart from the left ventricle were arteries (originating from the heart), cf. Harris, 179.

47. Praxagoras fragment 11, Steckerl (1958).
48. Praxagoras fragments 7, 8, 9, 11, Steckerl (1958).
49. For more details on *pneuma* and pneumatics, see Berryman (2009): ch. 5.
50. Praxagoras fragments 5, 9, 84, 85; Phylotimus fragment 13, Steckerl (1958), cf. Galen, *De trem. palp. convuls. et rig.* 6, K VII 614.
51. Praxagoras fragment 99, Steckerl (1958).
52. Boylan (1982).
53. Steckerl (1958): 20.
54. Getting *pneuma* into the blood is a difficulty. This is because we have no account in Praxagoras of a cross-over between the arterial and venous systems.
55. Praxagoras fragment 32, Steckerl (1958).
56. Galen, *De tremore, palpitatione, convulsion, et rigore*, 1, K VII: 584.
57. However, the manner of relation is not set out. It would certainly make a big difference whether we are talking about an antecedent causal condition or a consequence of some other process.
58. Galen, *De differentiis pulsuum* 4, K VIII: 723.
59. Praxagoras fragment 11, Steckerl (1958) from Galen *De placitis Hippocratis et Platonis libri novem*. I. Müller, ed. (Amsterdam: A. M. Hakkert, 1975): I, 1, p. 143, see also DeLacy, 1981.
60. This book will recognize as presently authoritative for Herophilus: H. von Staden, ed. *Herophilus: The Art of Medicine in Early Alexandria* (Cambridge: Cambridge University Press, 1989) and for Erasistratus: Ivan Garofalo, ed. *Erasistrati Fragmenta* (Pisa: Giardini, 1988).
61. Praxagoras fragment 3, Steckerl (1958).
62. Celsus, *On Medicine*. 3 vols. ed. and trans. W. G. Spencer (Cambridge: Harvard University Press (Loeb), 1935): Prooemium 8.
63. Ibid, 14–18.
64. Ibid, 14–25. John Scarborough, "Celsus on Human Vivisection at Ptolemaic Alexandria" *Clio Medica* 11 (1976): 25–38, argues that Celsus may have attributed dissection and vivisection to the Ptolemies as a way of showing them to be barbaric, that is, a negative polemic (given the strong Greek taboos against defiling a human body). Lloyd (1973): 77, sets out independently that the Ptolemies permitted vivisection on condemned criminals. I agree with von Staden: 145 that since Celsus is a defender of Dogmatic vivisection (which included Herophilus and Erasistratus), and that he is not generally inclined to political polemics, it is probable that they performed both dissection and vivisection.
65. Von Staden (1989): ch. 5.
66. For a general discussion of this problem, see Boylan (2008): 106–114.
67. There are many instances of this, such as Achilles' tawdry treatment of Hector (*Iliad*) to Creon's treatment of Polynices and Antigone's heroic response for her brother's body (Sophocles, *Antigone*).
68. Certainly this side benefit is generally for the sake of educating medical students, who are present as observers, and surgical residents, who are present to learn surgical techniques that depend upon knowledge of anatomy and physiology.
69. I explore the ethical implications of vivisection in this very context in Michael Boylan, "Ethical Limitations on Scientific Research" in *Scientific Freedom*, eds. Simona Giordano, John Coggon, and Marco Cappato (London: Bloomsbury, 2012): 149–161.
70. For a riveting account of this procedure as a device to get people to agree with his theory, see Thomas Wright, *Circulation: William Harvey's Revolutionary Idea* (London: Vintage, 2013): 127–131. It is Wright's hypothesis that this dramatic presentation was essential in getting his audience to at least consider the heart as a pump and blood circulation as a possibility.

71. Arrian; vol. 1: *Anabasis Alexandri*, books i–iv; vol. 2: *Anabasis Alexandri*, books v–vii. Ed. and tr. P. A. Brunt (Cambridge: Cambridge University Press, 1976): 3.1, 1–3; 2.2 document Alexander of Macedon's role in the founding and marking out of the city. For a popular account of what happened to Alexander's body and the military/political milieu in which this occurred, see James Romm, *Ghost on the Throne: The Death of Alexander the Great and the Bloody Fight for His Empire* (New York: Vintage, 2012).
72. Nutton (2013): 132–133.
73. In Herodotus 3.1 and 3.129 Egyptian physicians were sought even to Persia (Herodoti *Historiae*, ed. Carolus Hude (Oxford: Oxford University Press, 1927), cf. A. S. Yahuda, "Medical and Anatomical Terms in the Pentateuch in Light of Egyptian Medical Papyri," *Journal of the History of Medicine and Allied Sciences* 2 (1947): 549–574. There is also the famous case of an Egyptian drug given to Helen of Troy in the *Odyssey* IV: 220–232 that would render the patient able to accept severe mental trauma for a couple of days. This *pharmacon* was well known in the ancient world: Pliny, *Natural History*, ed. and tr. H. Rackham (Cambridge, MA: Harvard, 1944): XXV.5, 11–12.
74. Herodotus (Herodoti), II.77, 3; II.84. See also, Jacques Jouanna (2012-b) on this very episode along with the role of wine as a mixing agent for Greek medicine.
75. This is not to suggest that the Greeks and Egyptians shared the same magical vehicles. Rather, the point to be made is that they both *employed* magic alongside *phusis*$_1$ medical practice.
76. The "London Papyrus" (British Museum 10059, 37, cf. 158).
77. Hermann Grapow, ed. et al. *Grundriss der Medizin der alten Ägypter* 9 vol. (Berlin: Akademie Verlag, 1954–1973): IV, 64; cf. II, 11–26 on "Zaubersprüche."
78. E. Thrämer, "Asklepios (2)" *Pauly-Wissowas Realencyclopädia dell'arte antica classica e orientale* 2 vols. (Stuttgart: J.B. Metzler, 1894): 1642–1697; cites more than 180 sanctuaries of Asclepius, but Thrämer in "Health and Gods of Healing (Greek)" *Encyclopedia of Religion and Ethics* 6 (1914): 540–553 suggests the number may be as high as over 400 existent ruins. However, the number at the time was most likely much greater.
79. Herophilus fragment T248c, von Staden.
80. Jacques Jouanna (2012a) discusses a number of influences upon Greek medicine from Egypt.
81. For a discussion of the conditions that permitted this radical change in Alexandria—particularly concerning dissection and vivisection, see Heinrich von Staden, *Herophilus: The Art of Medicine in Early Alexandria: translation and essays* (Cambridge: Cambridge University Press, 1989), cf. von Staden. "The Discovery of the Body: Human Dissection and its Cultural Contexts in Ancient Greece" *The Yale Journal of Biology and Medicine* 68 (1992): 223–241. The distance traveled between the reverence for the corpse in *Antigone* to the permissible cutting of skin and organs cannot be minimized.
82. *CMÄ* VI; 4.1, 2: 44–81; cf. Geoffrey Lloyd (1987): ch. 5 who argues that measurements, in general, by the Egyptians were rather loose and quite inaccurate.
83. There are a few possible questions about whether Erasistratus was physically in Alexandria. See: P. M. Fraser, "The Career of Erasistratus of Ceos," *Rendiconti*, Classe di Lettere Instituto Lombardy—Academia de Scienze e Lettere, 103 (Milan, 1969): 518–537. Basically, Fraser expands on Wellmann's queries. Also: Nutton (2013): 135.
84. Galen, *De usu partium*. VI.10, von Staden, Fragment T 116, cf. T 117 Rufus of Ephesus that calls into question the classification of the pulmonary vein.

85. Galen, *De usu partium* VI.10, K. III, 445.
86. Bilateral symmetry *can* be viewed in different ways. On the one hand, the origins of the maxim of bilateral symmetry as a true component of humans (and other living organisms) comes from observation, which is a cornerstone of *phusis*$_1$. However, when this is generalized beyond what can be observably corroborated, then it becomes an unsubstantiated instance of faith, *phusis*$_2$. I am suggesting that in this case Praxagoras was in the latter position and Herophilus follows his lead.
87. Von Staden (1989): 260–267; Longrigg (1993): 200–202 believes that Herophilus stuck with his teacher's *pneuma*-alone theory.
88. Harris: 178, cf. Herophilus fragment T 79, von Staden on work on the brain.
89. Herophilus fragment T79, 122–124, 90–92, 98–99, von Staden.
90. Herophilus fragment T121, von Staden.
91. Herophilus fragment T 80, von Staden.
92. Herophilus fragment T 80–85, von Staden.
93. Galen, *De diff. puls*. IV.2, K. V: 561.
94. Harris, 183–195.
95. Erasistratus fragments 195–196, Garofalo.
96. Erasistratus fragment 201, Garofalo.
97. Strictly speaking, a relaxing of the heart muscle (modern), but here, fluid comes in, and in part, some of this dynamic is just as Erasistratus suggests.
98. Wright, essay 3.
99. Erasistratus fragment 47–49 Garofalo, cf. Harris, 195–233.
100. Cf. Garofalo: 33–43.
101. This sort of experimentation consists of creating a single perturbation in the natural order and observing what happens. This sort of experimentation is very different from the modern hypothetical-deductive method in which there are control groups to form single- or double-blind constraints based upon dependent and independent variables so that the possibility of error in interpretation is greatly reduced. This is because it is set out in the beginning where the terms of success or failure are structured into the process. For a quick primer on modern scientific method, see Steven Gimbel, *Exploring the Scientific Method: Cases and Questions* (Chicago: University of Chicago Press, 2011). For an example of the ancient notion of "experiment" set in the contemporary world, see the renowned ethnologist Konrad Lorenz, *On Aggression*. 2nd ed. (London: Routledge, 1996). The ethnologists believe that the creation of control groups within the context of an isolated target creates possible sources of error through the artificiality of the method.
102. Again, this is a perennial problem in the philosophy of science. The most contemporary scientific school that employs the single perturbation method is the European ethological school. They study animal behavior by altering one structure of the natural environment and observing what happens. An emblematic practitioner is Konrad Lorenz, *King Solomon's Ring*. 2nd ed. (London: Routledge, 2002), cf. Lorenz (1996). Of course, a more nuanced approach involves statistical induction following John Stuart Mill's work (1843). This includes adding other variables into the problem-solving process. For example, though the so-called methods of agreement and difference are both single perturbation accounts, the joint method of agreement and difference adds one more level of analysis (cf. the Zeeman effect that examines magnetic fields in the presence of an open oxyhydrogen flame to describe the alterations of the sodium spectrum). Other multifactorial statistical models include the method of residues (with the classic case being

the discovery of Neptune) and the method of concomitant variation (with the classic case being the description of the tides). For other statistical analysis along this line, see Ian Hacking (2001). Of course, Popper and his followers will demur here.

103. *De anat. Admin.* VII. 16; K. II: 646–648; cf. discussion by Harris, 377–386.
104. *De sang in art.* 8; K. IV: 733–735.
105. Vesalius, *De humani corporis fabrica*, VII, cap. 19.
106. William Harvey (1628) *Exercitatio Anatomica de Motu Cordis et Anguinis in Animalibus*. Frankfurt: W. Fitzer/rpt. London: Sydenham Society): 111; cf. comments by J. M. Forrester (1954) "An Experiment of Galen's Repeated" *Proceedings of the Royal Society of Medicine* 47.4: 241–244. An interesting twist on this was suggested to me by Nutton that Harvey carried out hundreds of animal experiments, and his notion of blood flow was helped by working on cold-blooded animals where the flow is much slower and movements easier to observe.
107. Hacking: chs. 14–15.
108. Hume's famous exposition of this position is as follows:

    1. There are two sorts of skepticism: extreme and mitigated—Assertion (p. 111)
    2. Extreme skepticism arises out of paradoxes contained within abstract reasoning (e.g., infinity, Berkeley, Pyrrhic)—Assertion (p. 108)
    3. The extreme skeptic cannot answer what his meaning is [the self-referential question]—Fact (p. 110)
    4. [Whatever cannot give an account of his meaning should not be believed]—Assertion
    5. Extreme skepticism should be rejected—1–4 (p. 110)
    6. Mitigated skepticism calls possible errors into question regarding the limits of knowledge—Assertion (p. 111)
    7. [Freeing oneself from error is good]—Fact
    8. Only mitigated skepticism should be adopted—1, 5, 6, 7 (p. 112)
    9. The two sorts of knowledge are matters of fact and matters of reason [synthetic and analytic knowledge]—Fact (p. 112)
    10. [Analytic statements demonstrate within a closed system]—Fact (p. 112)
    11. [Closed systems can prove anything (depending upon their conventional rules)]—Fact
    12. Analytic statements can prove anything—10, 11 (p. 113)
    13. [Analytic statements alone resemble the Medieval Schoolman Metaphysics]—Fact
    14. [Whatever can prove anything should not be believed]—Fact

    ____

    15. Knowledge should begin with mitigated skepticism and proceed using both matters of fact and matters of reason; reason alone should not be believed and its books should be committed to the flames—8, 12–14 (p. 114) David Hume, *An Enquiry Concerning Human Understanding*. Eric Steinbert, ed. (Indianapolis, IN: Hackett: 1977 [1748]). Bracketed premises are enthymemes.

109. Meditation II (Charles Adams and Paul Tannery, *Oeuvres de Descartes* 11 vols. (Paris: J. Vrin, 1983 [1897]): VII, 25–28)
    Deceiver or not, I exist (A&T VII, 25–28):

    1. An evil deceiver tries to trick me or not—Fact
    2. [I] think or "thinking is"—Fact
    3. [All verbs require subjects and objects]—Fact

4. *I* exist without a deceiver—2, 3
5. If there is a deceiver he tries to trick *me*—Assertion
6. I exist with a deceiver—3, 5
7. Deceiver or not I exist—1, 4, 6

110. For a brief introduction to this enormous literature, see Georges Dicker, *Descartes: An Analytical and Historical Introduction*. 2nd ed. (New York: Oxford University Press, 2013); Willis Doney, ed. *Eternal Truths and The Cartesian Circle: A Collection of Studies* (New York: Garland, 1987), cf. one discussion thread that is very prominent: Alan Gewirth, "The Cartesian Circle" *Philosophical Review* 50 (1941): 368–395; Alan Gewirth, "The Cartesian Circle Reconsidered" *The Journal of Philosophy* 67 (1970): 668–685. An example of a supporter of Descartes within the context of the European medieval philosophy (seen as an offshoot of ancient Greek philosophy) is Andreas Hüttemann, "Descartes' Kritik an den realen Qualitäten: das Beispiel der Schwere" *Archiv für Geschichte der Philosophie* 83 (2000): 24–44.
111. Whether one can get to scientific realism via radical doubt is much disputed. This is because a moment of epistemic closure is required, and this seems to go against the impulse of radical doubt. For an exploration of some of these issues, see Sven Rosenkranz, "Radical Scepticism Without Epistemic Closure" *Philosophy and Phenomenological Research* 85.3 (2012): 692–718.
112. Boylan (2008): ch. 4; cf. William Kingdom Clifford, (1901). "
113. Σκεψις, on empirical consideration, Plato, *Phaedrus* 83a; on continued speculation Hippocrates *VM* 4, Plato, *Theaetetus, 201a, Phaedrus* 237d, Aristotle *Politics* 1254a 34. One recent study relates this intentional lack of closure to epistemological relativism. See Howard Sankey, "Scepticism, Relativism and the Argument from the Criterion" *Studies in History and Philosophy of Science* 43 (2012): 182–190.
114. Aristotle, *Physics* 239b9 (DK 29A 25); for a discussion of expanded interpretations beyond the mere mathematical puzzle, see David Sherry, "Zeno's Metrical Paradox Revisited" *Philosophy of Science* 55 (1988): 58–73 and Alba Papa-Grimaldi, "Why Mathematical Solutions to Zeno's Paradoxes Miss the Point: Zeno's One and Many Relation and Parmenides' Problem" *Review of Metaphysics* 50.2 (1996): 299–314.
115. The role of *logos* is set out in: "Listening not to me but to the Logos it is wise to agree that all things are one" KR 199, cf. 197. The ground of the skepticism is the balancing of opposites: KR 203 "The path up and down are one and the same." The famous stream metaphor is: "Upon those that step into the same rivers different and different waters flow. . . . It scatters and . . . gathers . . . it comes together and flows away . . . approaches and departs" (KR 217).
116. Another modern proponent of this sort of epistemological sortie is Bishop Berkeley, mono-non-materialism, *On the Principles of Human Knowledge* (Dublin: A. Rhames for J. Pepyat, 1710): sections 3, 6, 22, 24, 33.
117. Boylan (2008): 71–77, cf. William James, "The Sentiment of Rationality" in *The Will to Believe and Other Essays* (New York: Longmans, Green, & Co. 1897): 63–110.
118. For a thorough examination of these authors for whom texts are scarce, see R.J. Hankinson, *The Skeptics: The Arguments of the Philosophers* (London: Routledge, 1995).
119. Sextus Empiricus, *Outlines of Pyrrhonism*, tr. R. G. Bury (Cambridge, MA: Harvard University Press, 1933): I.4.
120. *Ibid.*, I.VI.
121. *Ibid.* I.VII.
122. The so-called self-referential paradox would be that if the skeptic says that no theories can be accepted and the opposite is just as true, and if skepticism is a

theory, then Dogmatism (the opposite) is just as true as skepticism. Augustine tries to make a similar point against the skeptics in the following argument:

1. "P or ~p" cannot be denied—F
2. That which cannot be denied is true—A

---

3. Either p or ~p is true—1, 2

What Augustine is doing here is to contend that the law of the excluded middle underlies all utterances. Thus, to somehow contend that both "p" and "~p" can be held together offends reason, and that any enunciation—such as the definition of skepticism—requires that one also admit that skepticism, itself, must be true or false (p or ~p). See Augustine, *Against the Academicians*, Bk II, vols. 11–12; Bk III, vols. 23–26, ed. Mary Garvey (Milwaukee, WI: Marquette University Press, 1942).

123. For a discussion of this problem, see Mauri Valtonen and Hannu Karttunen (2006).
124. Sometimes this is viewed as an instance of what should be known. For an examination of this, see Michael Boylan, "Ethical Limitations on Scientific Research" in *Scientific Freedom*, ed. Simonna Giordano, John Coggon, and Marco Cappato (London: Bloomsbury, 2012): 149–161.
125. Readers are referred to the following to make sense of this chart: (1) Three senses of *phusis* (ch. 1); (2) Necessity scale from D-□ to D-◊ to E-◊ (ch. 2); (3) Type-token distinction (ch. 2): (4) Explanation-type α and β (ch. 2).
126. I give an overview of epistemological externalism in Boylan (2008): ch. 4.
127. In modern French, this duality is maintained in "experience," which can refer to the uncritical (naïve) single perturbation style of experiment, along with the more critical reflections on such: the hypothetical-deductive method.
128. In support of this approach, see Karl Deichgräber, *Die griechische Empirikerschule*, 2nd ed. (Berlin: Weidmann: 1965); Michael Frede, "The Ancient Empiricists" in *Essays in Ancient Philosophy* (Oxford: Clarendon Press, 1987): 243–260; R. J. Hankinson, "The Growth of Medical Empiricism" in D. Bates, ed. *Knowledge and the Scholarly Medical Traditions* (Cambridge: Cambridge University Press, 1995): 41–59; and Nutton (2013). For a contrary opinion, see Hankinson (1995b): 226. Another contrary view pushes the date much earlier; see Edelstein (1967): 195–203.
129. Some of these heavy epistemological requirements are set out by M. Vegetti, "L'Épistémologie d'Érasistrate et la Technologie Hellénistique" *Ancient Medicine in Its Socio-Cultural Context; Papers Read at the Congress Held at Leiden University, 13–15 April 1992*. 28 (1995): 461–472.
130. Von Staden (1992): 235. On criticisms of folk psychology along this and affiliated arguments, see Paul Churchland, "Eliminative Materialism and the Propositional Attitudes" *Journal of Philosophy* 78.2 (1981): 67–90; and Jerry A. Fodor, "The Mind-Body Problem" *Scientific American* 244 (1981): 114–123.
131. Von Staden (1992): 236.
132. Deichgräber (1965): 83, 165 describes Glaucias' work *The Tripod* that sets out this doctrine.
133. Harris (1973): 193.
134. Of course, evidence-based medicine is still in combat with theory-based medicine. See, for example, the following exchange in the *British Medical Journal*. F. Davidoff, B. Hayne, D. Sackett, R. Smith, "Evidence-Based Medicine" *BMJ* 310 (1995): 1085–1086; and a reply in the letters section—N. J. Pearson, J. Sarangi, R. Fay, "Evidence-Based Medicine" *BMJ* 312 (1996): 380; and finally David Sackett, William M. C. Rosenbert, J. A. Muir Gray, R. Brian Haynes, W. Scott Richardson "Evidence-Based Medicine:

What It Is and What It Isn't" *BMJ* 312 (1996): 71–72. For a version of this based upon the Empiricists vs. Dogmatists, see Mohan Matthen, "Empiricism and Ontology in Ancient Medicine" *Apeiron* 21 (1988): 99–121. Matthen claims that because Empiricists are not committed to logically based theories that may belie any particular instance of falsification, the Empiricists were really in a better position to jettison treatments that simply did not work. This is because they did not have a general theory that they have bought into. This is essentially the same dynamic in the modern context.

135. Of course, reference to humors is not a "stand-alone" position. But since cutting that is more invasive will entail the loss of considerable blood, surgeons who do not take this into account will be largely unsuccessful. This is why I suggest that surgeons who attempt more invasive techniques will be inclined to the Dogmatic approach.
136. These extend before the Hippocratic Writers and include the accumulation of written texts/short guides, summaries, etc.
137. Two contemporary examples of folk medicine practiced by women within a family/social setting are Jewel Babb and Pat Littledog, *Border Healing Woman: The Story of Jewel Babb as told to Pat Littledog*. 2nd ed. (Austin: University of Texas Press, 1999) and Michael Mucz, *Baba's Kitchen Medicines* (Edmonton: University of Alberta Press, 2011). Another interesting case concerns modern malaria treatments that embrace folk herbal medicine in Africa in what is termed "reverse pharmacology" in which folk herbal treatments are used—such as *Argemone mexicana* tea for malaria in Mali (that has an 89% recovery rate—compared to 95% with ACT, artemisinin combination therapy); see Bertrand Graz, et al. (January, 2010) "Argemone Mexicana Decoction versus Artesunate—Artesunate-Amodiaquine for the Management of Malaria I Mali: Policy and Public Health Implications" *Transactions of the Royal Society of Tropical Medicine and Hygiene*. 104.1: 33–41; and Merlin Wilcox (March, 2012) "Improved Traditional Medicines in Mali" *Journal of Alternative and Complementary Medicine*. 18.3: 212–220.
138. For a more complete account, see Boylan (2008): 78–84.
139. Of course, the words *do no harm* do not occur in the Oath, but have come to be the common interpretation of "I will help the sick according to my skill and judgment, but never with an intent to do harm or injury to another" (ἐπὶ δηλήσει δὲ καὶ ἀδικίη εἴρξειὐ), ll. 18–19)—see Boylan's translation (2014b).
140. Galen, "An Outline of Empiricism" in *Three Treatises on the Nature of Science*, tr. R. Walzer and M. Frede (Indianapolis, IN: Hackett, 1985): chs. 2, 6, cf. Hankinson (1995b): 227–228.
141. Cf. David Hume's argument on this exact point (*Enquiry*—1977 [1748]):

    1. "Causation" implies non-logical connection between cause and effect—Fact (p. 49)
    2. In matters of fact sentiments are formed through direct experience—Fact (p. 49)
    3. When two sentiments are often linked together the ideas that are formed from them will also be linked—Assertion (p. 50)
    4. The constant conjunction of objects will give rise to an idea of the objects being linked—2, 3 (p. 51)
    5. There is no necessary relation between these two objects—Fact (p. 51)
    6. The two habitually joined objects are not, strictly speaking, cause and effect—1, 5 (p. 51)
    7. [What is not necessary is contingent]—Fact

    ---

    8. What is commonly called cause and effect is really two contingently linked objects and not properly called causation—4, 6, 7 (p. 52)

142. For a discussion of this in a general scientific context, see Elizabeth Asmis, *Epicurus' Scientific Method* (Ithaca, NY: Cornell University Press, 1984).
143. Galen, *An Outline of* Empiricism Frede and Walzer, tr. (1985), ch. 8.
144. Galen, Fre *On Medical Experience* Frede and Walzer, tr. (1985): chs. 3–6.
145. Galen, K: 12, 534, 989; 13, 462.
146. Nutton (2013): 152.
147. Most take the school of the method to have been founded by Themison of Laodicea, a pupil of Asclepiades of Bithynia (cf. Wellmann (1922): 396, 401, 405). Then Thessalos put everything together around the time of Nero, adding the doctrine of *metasynkrisis*, on modifying the pores of the body toward tightness or looseness, and the doctrine of three-day intervals for evaluating disease, *diatritos* (cf. Nutton (2013): 197). Since there are few texts, this book will follow this account. It should be noted that Edelstein very strenuously denies that Themison be viewed as the founder of the method—leaving that role to Thessalos (Edelstein 1967): 174–179.
148. Edelstein (1967): 181 calls these moist and dry, and they affect the whole body. It should be stated that the Methodists were not keen on strict definitions. This is because that approach characterized the Dogmatists, against whom they sought professional differentiation.
149. Nutton (2013): 195.
150. D. Leith (2008) "The *diatritus* and Therapy in Graeco-Roman Medicine" *Classical Quarterly 58*: 58–600.
151. Pseudo-Galen, Kühn XIV, 684.
152. For a discussion on how the various theories of truth models work, see Boylan (2009): ch. 4.
153. Though Soranus said that reading natural philosophy might be a good discipline for the mind in the form of juvenile education, this theoretical part is " useless for our purpose" (Soranus, *The Gynecology*, tr. Temkin): 4.
154. There are, of course, the fragments and testimonials collected by Tecusan (2004)—which are substantial. However, I also think that there is a non-theoretical-skeptical bent to the Methodists that give a reason for why Galen became dominant. My analysis goes against Edelstein's analysis of the skeptical connection with the Methodists: Edelstein (1967): 186–189.
155. As per Demont and Jouanna (1981): 197–209.
156. This does not totally dismiss the magical *phusis*$_2$ sense. For example, after this very passage, Soranus brings up the use of magical amulets, saying that though he does not use them, he doesn't forbid them either: "Yet one should not forbid their use; for even if the amulet has no direct effect, still through hope it will possibly make the patient more cheerful" (Soranus III.10.42).
157. Soranus, III, 10.41
158. Soranus rejects bloodletting (contra Themison) because it does not contract, but further loosens the blood flow and can therefore threaten the life of the woman, Soranus III. 10. 42. For more on this, see Chapter 5, this volume.
159. Galen, (*On the Sects for Beginners*, Walzer, and Frede tr. (1985) ch. 6, and Celsus, *De Medicina*, "Prooemium" 62–73. R.J. Hankinson argues in "Causes and Empiricism" *Phronesis* 32.3 (1987): 329–348 that the "hidden" was confronted by the Empiricists via inference from common observation and by associative signs—for example, if a woman is lactating, then she has conceived. The empirically verifiable lactation is a sign of pregnancy. One may infer the latter from the former—though, strictly speaking, it is not an instance of causation (à la Hume), cf. David Sedley, "On Signs" in Jonathan Barnes, et al. eds. *Science and Speculation* (Cambridge: Cambridge University Press, 1982): 239–266.

5 Galen—The Grand Synthesis

Our express tour of ancient biomedical writers is about to end. We have made many philosophical stops along the way in order that we might better appreciate our terminus when we get there. That moment is now approaching. In Galen we have a writer who thoroughly takes into account those who came before him and tries to glean the best from them all (even those with whom he generally disagrees).[1] This is the methodology of a synthetic, systematic philosopher.[2]

The synthetic philosopher tries to unite the leading positions of those who came before him. This includes looking at those with whom one is sympathetic—though perhaps for not the same reason that the predecessor set out. Those against whom the synthetic philosopher disagrees constitute an opportunity to further his own position by opposing theoretical or clinical tenets that are put forth.

Thus, the synthetic practitioner sees himself in a context. He is not inventing everything. He recognizes his position in an ongoing tradition. But not all philosophers/biomedical practitioners contextualize this. Many of the pre-Socratic philosophers and Hippocratic Writers set out their positions as if they were thinking about these issues for the first time. When they do make references to others, they are rather oblique. These philosophers believe that they create a totally novel theory *out of whole cloth*. This distinction refers to the nature of the philosopher's project—contextualized (synthetic) or completely novel.

A second distinction is the aspiration of the presentation rendered. Many synthetic philosophers are drawn to a *systematic* account that seeks to connect the entire *explanandum* in some way. In modern biology, the theory of evolution serves this function. In the ancient world, there were several candidates: humor theory, the contraries (hot, cold, wet, dry), somatic symmetry, "right side" over "left side," teleology (Nature does nothing in vain), etc. Combining these broad explanatory devices employs the Dogmatic ("D") approach and seeks modal certainty, $\Box$, or at least a firm contingency that is generally certain ("D-$\Diamond$"). The ontology is generally *phusis*$_1$, which yields a skepticism$_2$ through an epistemology of the careful scrutiny maxim and explanatory-type α as it considers tokens and types equally.[3]

The complement to the *systematic* philosopher is the *eristic* philosopher. Eristic practitioners are interested in solving the particular issues before them (the clinical approach). They are focused upon the token. Their ontological interest is in *phusis*$_3$. They tend toward skepticism$_3$ and adopt an extreme version of the epistemological careful scrutiny maxim. The Empiricists and Methodists fall into this category. Many eristic philosophers also adopt the *whole-cloth* approach because they eschew *historia* except as a clinical guidebook of what has worked in the past.

Plato stands as an eristic philosopher who only sets out other positions as a pretext to defeating them.[4] He puts the positions that are wrong into the mouths of characters who make gross mistakes in logic.

The Methodists and Empiricists stand in this tradition of only looking at the case at hand without regard for theoretical arguments of others (by principle). However, they do accept handbooks of symptoms and treatments by earlier thinkers without comment.

The synthetic philosopher is Dogmatist by definition. He accepts *arguments* and *depictions* that have come before. His task is to think about the claim at hand and then to give a better *explanans* than anyone previously.

The systematic philosopher is one who imagines an integrated whole in which everything fits coherently together. Aristotle is one such dual thinker (synthetic and systematic), and Galen is another.[5]

What separates the synthetic, systematic thinker from either the eristic philosopher (who concentrates upon a specified problem in and of itself) or the disconnected systematic thinker (who imagines a totally new model of expression) is that in the case of the eristic thinker particular problems drive the discussion without a need to cite anyone else (e.g., Plato, Descartes, Hume). In the case of the whole-cloth thinker, one creates an entire interrelated universe out of posits that are asserted to have come only from the philosopher.

What is important here is that the eristic, nonsystematic philosopher is concerned only with particular problems. Followers of these thinkers often create their own systematic account. Often, these commentators' evaluations are so accepted that many forget that the systematic step came from the commentator and not from the original author—such as Plato having or not having a "Theory of Forms."[6]

This stand-alone systematic project is generally discounted in philosophy, though there are some isolated exceptions that have many advocates (e.g., Leibnitz, Hegel, Whitehead). Thus, there is a mixed reaction to the disconnected (i.e., nonsynthetic), systematic philosopher. However, in the history of science there have been some notable champions of this form, for example, Galileo, Newton, and Einstein.[7]

This chapter will concentrate particularly upon Galen's work on the blood, heart, and pulse as it seeks to situate him within the intellectual explanatory itinerary that has been put forward in this book (see Chapter 4; Table 4.1, this volume). The first stop at our terminus is how Galen explicitly decides to discuss various tenets of his predecessors.

GALEN AND EARLIER WRITERS

If we are to understand Galen as a synthetic thinker who fashions a new systematic amalgam from those who came before along with his own original insights, it is important to review how Galen saw his intellectual context.[8] For Galen, there were two groups of writers to which he most often refers: early-fourth-century BCE writers and later biomedical writers. This subsection will concentrate upon the first group of fourth-century Greek natural philosophers such as Plato and Aristotle. The Hippocratic Writers also fall into this general group. The second group of writers is treated (with many citations from Galen) in Chapter 4 in this volume.

Galen picks and chooses what he finds useful from Plato and Aristotle. For example, Galen is attracted to the Platonic tripartite soul: *thumos* (spiritedness), *epithumos* (directed spiritedness, i.e., desire), and *sophia* (wisdom) via *logistikon* (rational analysis).[9]

From Aristotle, Galen gets a keen sense of critical Empiricism. This can be understood through Aristotle's deft utilization of various principles, such as the four elements (earth, air, fire, and water), along with the contraries (hot, cold, wet, and dry), in order to explain physical observations (see Chapter 3, this volume). These can be viewed as ontological simples that are various ranked, for example four levels of hot from mildest heat to scalding, as well as in composition—particularly in the use as a procedure for mixing drugs—that had to account for the properties of the drug and their complementary functions.[10]

Sometimes this utilization of various explanatory principles occurs for the sake of accounting for the function of parts. For example, it was just stated that there are three parts to the soul (from Plato), and it was generally agreed that there were three principal organs in the body (the brain, the heart, and the liver).[11] If one is interested in connecting parts to function, what better way than to connect the three principal parts of the soul (here understood as what later would be called vital functionalism) with the three principal organs? Each part of the soul finds a physical location in a bodily organ: *thumos* originates in the heart, *epithumos* in the liver, and *sophia/logistikon* in the brain. The three parts of the soul correspond to the three principal organs of the body.[12]

Behind this inclination to connect parts and functions is the notion that Nature does nothing in vain (*ouden he phusis ergazetai maten*).[13] This means that Nature would never allow a state of affairs to exist in which a body part had no function (subject to the "more and less"). All parts have functions (generally only one), and all functions are related to body parts. This principle of efficiency and simplicity guided Aristotle's biological science and, by extension, Galen's.

In turn, Galen makes a further link to a mechanical structure of sorts[14] by connecting the principal organs to delivery mechanisms. He does this primarily through observation of simian dissection.[15] The spinal cord seems to begin at the brain.[16] Therefore, the large nerves associated with the spinal cord may be particularly suited for doing what the brain is meant to do: govern thinking (*sophia, phronesis, theoretikon*, et al.). The material medium is *pneuma* (in a post-Aristotelian sense), so it seemed reasonable to Galen that psychic *pneuma* came to be in the brain and was communicated to the rest of the body via the nerves.[17] In a similar fashion, arteries are chosen for the heart because of the large blood vessels connected to the heart.[18] Thus, natural *pneuma* is created in the heart and is then disbursed to the body for its *thumos* function. Finally, because of the large hepatic venal structure about the liver, Galen thought that veins began there and transported desire to the rest of the body.[19] Therefore, one way that Galen followed Aristotle was in matching key functions to organs via an explanation-type α account. This leads to the second way that Galen followed Aristotle—making sure that the mechanical/material account described the activity *causally*.

As mentioned in Chapter 3 in this volume, Aristotle assembled four causes of natural phenomena: final cause, formal cause, material cause,[20] and efficient cause[21] in order to answer the four key questions of science: 1. *Hoti* (the fact), 2. *Dioti* (the reason for the fact), 3. *Ei esti* (if the fact is as such), and 4. *Ti esti* (the account of essence or definition; *Posterior Analytics* II.i.). Requiring a critical causal account of some natural phenomenon ensures that the matching that goes on between part and function is not arbitrary. In Aristotle, the four causes are often grouped together. Aristotle (at least in the biological works) invokes a strategy of redundant explanation (ME for the material/mechanical account, and TE for the teleological account).[22] Taken at its simplest level, he gives four accounts of everything. However, in actual practice, it comes about that Aristotle really only offers two accounts: the material/kinetic account, ME, and the essence/ purpose account, TE (see Chapter 3, this volume).[23]

Galen, however, adapted Aristotle to Galen's own purpose. For example, the final cause is sometimes distinguished in its goal-oriented function as something that needs attending to (*skopos*) and at other times it is an end to which all things strive, *telos*.[24] Galen also introduces the instrumental cause by which the tools necessary to complete some task are added to the picture.[25] In each of these instances Galen offers a more action-oriented understanding of cause (*aition*) than does Aristotle. It has been argued by Michael Frede (1980) that earlier causal accounts (such as Aristotle's) merely offered the backdrop against which change occurred. This understanding of cause transformed (as a result of the Stoics in later Hellenistic times) to more active agents in bringing about outcomes.[26] However, as Chapter 3 in this volume points out, I argue for a more robust account of Aristotle's understanding of causation that is at least analogically connected to modern accounts of critical Empiricism.[27]

Galen picks and chooses what he wants to employ from Plato's philosophy and Aristotle's biological science. This synthetic activity is largely Galen's—though to be fair, his assessment of earlier writers may also have been influenced by the Stoics.[28]

Finally, there are the Hippocratic Writers. From the Hippocratic Writers, Galen received two things: (1) an appreciation of the unique empirical milieu surrounding sickness or accident of a given individual; and (2) a complicated set of ways to go about diagnosis and prognosis in order to prescribe treatment.

In the first case, one must give credit to the Hippocratic Writers for observation—though the way it is used varies. One of the Hippocratic Writers noted that disease is brought about by "breaths," *phusai*, which arise from undigested food.

> Breath (*pneuma*), he [Hippocrates] holds, is the most necessary and the supreme component in us, since health is the result of its free, and disease of its impeded passage. We in fact present a likeness to plants. For as they are rooted in the earth, so we too are rooted in the air, by our nostrils and by our whole body (K VI, 14, tr. Jones).

Air and *vital* air (*pneuma*) are related (see Chapter 2, this volume). The one is available to all. The latter has been processed in some way that is significant to humans. In this passage, it is clear that the writer is putting forth a very basic assumption about some primal principle of life, health, and disease. Galen cites this passage in the process of developing his own theory of *pneuma*.

Another important point that Galen attributes to the Hippocratic Writers is "skin breathing." This becomes an important point in the following way. For Galen, the breath is a source of air (and later *pneuma*), but if this is the *only* source of air, then if a sufficient quantity of air is needed to supply the body with its *pneuma* needs, and if the amount of air from the lungs is roughly equivalent in the inhale and exhale, then there must be another source of air to the body, viz., the pores in the skin.[29]

Galen supports the opinion of the Hippocratic Writer of *Epidemics* IV who says that the entire body breathes in and out.[30] Further, in *De usu respirationis* Galen asserts that arteries breathe through the skin, the brain through nostrils, and the heart via the lungs.[31] This may mean that, like his appropriations from Plato and Aristotle, Galen expands greatly a small doctrine of skin breathing for his own purposes.[32]

In the second case, one must address the Cos vs. Cnidos issue (discussed in Chapter 2, this volume). Though this classification is controversial, it is useful, whether one accepts the literal geographical demarcation, to mark some clear distinctions in the Hippocratic body of writing.[33] Galen, the synthesizer, takes the Dogmatist inclination[34] for D-◊ from some of the Hippocratic Writers (e.g., *The Nature of* Man) and the E-◊ sensibility from others (e.g., *Epidemics* II and III). This is not to say that Galen should be exactly placed in between (where we put the Methodists in Chapter 4, this volume), but rather he is a D-◊ thinker who does more than acknowledge the pesky *epi to polu* counter examples and reinforces the Dogmatist D-◊

critical Empiricist position with a keen appreciation for the individuality of the patient (clinical biomedicine).

In either case, Galen has a penchant for explanation-type α (see Chapter 2, this volume)—a sequential, mechanical causal process employing $phusis_1$ by emphasizing both types and tokens—that fits into the modality continuum at D◊ (see Chapter 2, this volume).[35]

Another key theoretical point that Galen adopts from the Hippocratic Writers is the notion of "attraction," *holke*.[36] This doctrine asserts a fluid transport mechanism. Since the heart is not a pump, there must be a material cause that accounts for blood and *pneuma* to move about the body.[37] Under this doctrine, an organ can exert the force of attraction to move fluids (like blood or any other humor)/*pneuma* to the part in question.

The most historically prominent theoretical scheme of the Coan writers was the doctrine of the four humors of the body: blood, phlegm, black bile, and yellow bile (or sometimes serum).[38] Health was defined as the balance of the four humors. Disease was defined as the imbalance of the humors. When imbalance occurs, then the physician might intervene by making a correction to bring the body back into balance. For example, if the individual were too full of phlegm (making her phlegmatic or lethargic), then the phlegm must be countered. Citrus fruit was thought by some to be a counteracting agent.[39] Thus, if one feels lethargic, increasing one's citrus intake will re-create balance. The treatment is, in fact, generally effective. Moderns might describe the therapy differently by ascribing the effect to vitamin C, phosphorus, and natural sugar. This example illustrates the scope of the Hippocratic physician in this context: something like a cross between the modern roles of an herbalist dietician and a personal trainer. Of course, whether a treatment *works* and a *cure* has been achieved is a highly subjective matter. Nonetheless, the cures that were dictated by the "four-humor theory" are reputed to have worked well enough for this theory to extend to the nineteenth century (in various guises).[40]

Other biomedical writers—some say from Cnidos and/or Sicily—held that strict empirical principles prohibited scientists from going beyond the data at hand. It was a better methodology for the biomedical practitioner to stay as close as possible to the clinical symptoms that were before him.[41] This meant that each patient would be seen in her particularity. Such a method required careful trial-and-error observation and only slight manipulation of the patient in the form of treatment. Galen accepted this to the extent that understanding individuality was necessary in order to know how to subsume these symptoms under some abstract rule. This, of course, is the form that contemporary evidence-based medicine follows (see Chapter 2, this volume).

There was a great conflict in the ancient world concerning the status of observational conclusions that were made by others, *historia* (see Chapter 4, this volume). Should they be given in their specificity and remain as disparate, individual accounts, or should they be grouped by similarity and more general principles drawn from them? In this instance, it was very much in dispute whether it was better to set out individual reports of particular illnesses (case studies) or to try to draw general rules from the particulars.

Take, for example, the Hippocratic work *Epidemics* III:

> If the gangrene mortifies itself there is a head pain and frequently a scratchy throat; the sick limb loses sensation, a feeling of cold comes to the head and the affected limb sweats. He suddenly loses his speech and blows blood from his nose as he becomes pale. If the disease takes hold of the patient with a weak force, he recovers the discharged blood. If the disease takes him with a strong force, he dies promptly. In this case one induces sneezing by pleasant substances; one evacuates by the upper and lower. Alternatively those odors will be a little active. The soup will be light and hot. Wine is absolutely forbidden (*Epidemics* III, Littré 7: 123).

In this passage one is left merely with symptoms and treatment. But when one practices medicine in this way, there are severe restrictions. For the disease is seen only as a collection of symptoms.[42] The cure can only be guessed at unless it has been previously written down in a manual. When a physician is confronted with a novel disease, he must find a similar set of symptoms and use that treatment. This aspect of the "trial and error" method brought harsh rebuke from Galen:

> The point is that they [the Cnidians and/or Sicilians] looked at the varieties of symptoms which change for many reasons and failed to consider the specificity of the dispositions, as did Hippocrates, who used for their discovery a method only by using which, one can find the number of diseases. . . . Hippocrates censures the Cnidian physicians for their ignorance of the genera and species of diseases, and he points out the divisions by which what seems to be one becomes many by being divided (*De Placitis Hippocratis et Platonis*, Iwanis Mueller 1874: 776, cf. Delacy, 1981).

What was it that made the Cnidians/Sicilians different from the Coan writers? This book has argued that the answer is rather complex and that the best answer is set out in Table 4.1 where most of the theoretical apparatuses are combined respecting the Dogmatists, Methodists, and Empiricists.

The effect of these differences can be found by examining the two steps in any medical response: prognosis and treatment. In the Coan work *On Prognosis*, the writer suggests that prognosis consists of knowing the patient's condition in the past, present, and future. Now how could a physician know this? Well, this could also have been part of a handbook catalogued through similar case studies. The practitioner could memorize each individual description. Next, the practitioner could add to this his own experience.

But the problem is that each case is individual. It possesses a unique set of properties.[43] The practitioner would not be in a good position to treat novel cases. When confronted with a novel case, the E-◊ practitioner is left with seeking *similar* cases. The implied premise is that similar cases call for similar remedies. The greater the experience, the more refined the practitioner can be in balancing similar cases with the remedies. As discussed in Chapter 4 in this volume, much rides on the word, 'similar.' If each individual presents in a unique way, then there is (at one level) no complete similarity. How much is enough? What counts? Physical symptoms? What if there are a bundle of physical symptoms for some particular disease? Which ones are essential and primary? Galen's version of Dogmatism can, in principle, be the more effective approach. This is because he emphasized the individuality of the patient along with a natural philosophical concern for the underlying hidden causes.

Such an alternative to the Empiricist program aims at establishing a theory of causes that underlie individual cases. These causes involve a depiction of *phusis*$_1$ elements such as earth, air, fire, and water, along with the contraries: the "hot," "cold," "wet," and "dry," or "the four humors." These elements are more general because they seek to describe a different level of analysis. They are type oriented (see Chapter 2, this volume). Diagnosis, prognosis, and treatment in this context refer to epistemological conditions that begin with observations of the individuality of actual patients, but are given meaning via an understanding of the hidden structure via *phusis*$_1$. This contextual meaning *situates* and *categorizes* various patient symptoms as being more or less important according to what they reveal about the hidden *according to the type-oriented theory*. Here we are interested in the genera and species of the disease in question. Such an exercise allows for a smoother transition between the patient and her particular symptoms (*tokens*) to *types* of diseases (see Chapter 2, this volume).

In order for this classification to avoid being based upon accidental characteristics, it is requisite that the classification includes the causal factors (as defined by the theory) that operate to bring about the disease in the first place.[44] This is really the *foundational* or *causal* network that is responsible for the disease's very existence. Such an understanding of the level of analysis moves away from individuals (tokens) and their "similarities" toward the theoretical (types). Mere connection may be accidental, but a causal framework makes the connection *real*.[45] Understood in this way, disease is an outcome of various hidden events within the patient that present according to a theory grounded in *phusis*$_1$. This systematic combination of token and type separates Galen from the Methodists, who view the disease as the primary entity. The hidden mechanism that caused the disease is not of interest.

The Hippocratic writings, as varied as they were, were influential in the development of later biomedical practitioners, including Galen's own systematic designs. The three principal Hellenistic schools—Dogmatists,

Methodists, and the Empiricists—all hearken back in various ways to the Hippocratic writings.[46] Many debates in the Hippocratic writings, such as the "preformation" vs. "epigenesis" debate, are picked up again and given a twist according to the predilections of the Hellenistic schools (see Chapter 3, this volume).[47]

GALEN—ON BLOOD, THE PULSE, AND THE ARTERIES

In the second part of this chapter we will speculate upon the *reasons why* Galen took the positions he did respecting blood and *the difference it makes.* Three questions will be addressed: (1) whether and why *pneuma* is mixed in with the blood; (2) why there is a pulse and how it operates, and (3) why the arteries are not merely pneumatic tubes and what roles might they play in the human organism.

A. Blood and *Pneuma*

The first question is whether and why *pneuma* coexists with the blood. Blood is one of the four humors of the body (along with yellow bile, black bile, and phlegm). Since blood is found in both veins and arteries, it is a vehicle for the other humors as they move about in the body. In this way blood is kind of master humor. Its material constitution would be affected by the four contraries: hot, cold, wet, and dry that made up the Empedoclean elements of earth, air, fire, and water.[48]

Following in the tradition of Aristotle, blood was essentially nourishment, *trophe*, to the body. It was transported to organs, sinews, and tissue and then "used up" as it was absorbed in the tissues.[49] Waste products were also carried in the blood (generally through the veins) to be sent into the intestines and kidneys.[50] Blood moves about, but does not circulate.[51]

What interests us most in this context is *aer*. As we have seen in Chapter 2 in this volume, *aer* can be related to the creation of *pneuma*. As noted earlier, there are two ways that *aer* enters the body: through the lungs and through the pores of the skin.[52] The exact details of this process are rather unclear. Here is a possible account of how the blood acquires *pneuma* given that it may contain *aer*. Since Galen, as a synthetic thinker (who is often just as busy with his reputation as he refutes Erasistratus or some other thinker), sometimes seems to contradict himself, this reconstruction is, by necessity, an imaginative exercise of what we say in modern science as filling in the data points by the process of curve averaging.[53]

First, let us consider the genesis of *pneuma*. There is an initial changing of air_1 received in the lungs and transferred by *anastomosis* to an artery.[54] Air in its common context is most often cool and dry (two of the four contraries). When the air from the lungs enters the artery, the air acquires innate heat and moisture such that it is transformed and becomes air_2. Air_2

is equivalent to an undifferentiated (pluripotential)[55] *pneuma*$_1$. In this state the *pneuma*$_1$ coexists (at least in the arteries, but probably also in the veins) with blood as a sort of potent mist. The declaration of seventeenth-century philosopher René Descartes that this force be termed "*spiritus animalis* or animal spirits" is perhaps less material than Galen would have liked. (Descartes in *Passions of the* Soul posited a dualistic metaphysics in which the nonmaterial soul has to gain a physical form in the pineal gland so that it might perform a vital function.)[56]

Nutton points out that unlike Aristotle, who posited a general perfect goal of balance between the contraries, Galen is happy to entertain various balances between both the contraries and the other humors.[57] Galen here begins his synthetic vision with the writer of *The Nature of Man* and his theory of the four humors. But Galen's own bent on the individuality of the patient caused him to posit not one but nine principal mixtures of contraries.[58] It might not be a bad thing for there to be many different examples of individual health. It was a judgment call on behalf of the physician.

In the case of the arteries, the blood is transported to the heart where *pneuma*$_1$ acquires an extra faculty, or *dunamis*, proper to the organ.[59] In the case of the heart, the new power is natural *pneuma* that is appropriate to the power of the soul accorded to the heart. (As mentioned earlier, the brain, heart, and liver all represent one aspect of the tripartite soul. The strict application from Plato here would make the power, *thumos*. However, it may be safe to interpret this as a general, natural power of life.)[60]

From the heart all this fully pneumatized blood is transported throughout the body. When some of this blood reaches the liver, a further *dunamis* is added to the *pneuma* transported in the veins. This next power is that of desire and is also shared with the humors producing black and yellow bile.[61] Thus, this sort of *pneuma* has three fellow travelers, blood, black bile, and yellow bile, via the venous system.

Finally, there is the *pneuma*$_2$ that reaches the brain. Here another change occurs in which the *pneuma* undergoes another transformation via the arterial system into psychic *pneuma*.[62]

In this account, we have followed air that has entered the lungs and traveled through the vascular system. But Galen was not satisfied with this account. For if breathing accounted for all of our *pneuma*, then how can we account for the fact that we exhale almost as much as we inhale? There had to be other sources as well. We have already discussed the breathing from the pores of the skin. In addition, psychic *pneuma* is created directly from some air$_1$ that is taken in through the nose and transported directly through bony cavities to the brain. This differentiates Galen's position from that of Erasistratus who held that the principal source of air to be the source material of *pneuma* was the lungs.[63]

Once in brain, the arterial blood penetrates the dura mater and divides into the famous rete mirabile delivering nourishment and *pneuma* to all parts of the brain.[64] Psychic *pneuma* is most like Cartesian *spiritus animalis*.

The power given is transported via the nerves and allows the brain to exert its will upon the movement of the body.

Thus, from the three major organs—brain, heart, and liver—the blood gains an increasing amount of *dunamis* so that it can perform specific functions. From the pluripotential particular, specification results in order to carry out the essential functions of life.

If this account is even close to being correct, then *pneuma* can be understood as a vital entity that fulfills various mechanical functions that we now ascribe to nutrition; natural vitality; assorted hormone-directed influences connected with desire, anger, quarrelsomeness; and generally all neurological activity. Since these various outcomes were of interest to the physician, they needed an account. When they got out of hand, they needed a remedy. Lacking a detailed explanation-type α account, Galen needed to have a rather open-ended entity/power/capability that might help bridge the gaps. Enter *pneuma*.

B. Pulse

The second question is why there is a pulse and how it operates. Let us address these questions in reverse order. When the physician touches the patient on the wrist, neck, or in the area of the groin, there is a discernable, rhythmic expansion of the arteries. If the arteries are filled only with *pneuma* (as Praxagoras and Erasistratus held), then this rhythmic expansion must be due to (a) a characteristic of the arteries themselves—they move up and down in an effort to support the motion of the *pneuma* or, in the case of Galen and Herophilus, the combined entities of blood/nourishment and *pneuma*; or (b) the *pneuma* itself; or (c) the *pneuma* as generated from some source (like the heart or lungs). Thus, before being able to address this problem, one must be clear what other theories are presupposed in order to form a background upon which to judge this question.

In this case, the *how* seems very important. If the arteries were the source of the pulse, then what would the purpose of this be? Option (a) suggests internal principles of the arteries. Now Galen, following Herophilus, accepted that the arteries had up to six additional covering layers.[65] This might be a start of making the pulse particular to this anatomical feature. But what would mechanically make the pulse happen, and how would the anatomical structure act to vary in pace? Doesn't this create a scenario in which a large number of complications arise? Since there is no single principle that one can refer to, it would seem that this sort of solution is, in principle, inscrutable. If one holds that the principles of science should be, in principle, open to inspection and discovery ($phusis_1$ and explanation type-α), then prima facie this alternative (a) must be rejected.

If it were the *pneuma* itself, then what would prompt the *pneuma* to expand at certain times and not others and to certain degrees of magnitude? As we have seen, *pneuma* possesses a rather plastic character that takes on various particular powers according to the context. Why should the context of coming from the heart make a pulse? Lacking an answer, one is inclined to reject (b).

This leads to the third alternative. Two possible sources are the lungs and the heart. At the beginning of *On the Use of the Pulse*, Galen asks, "What is the use of the pulse? Is it the same as that of breathing, as nearly all physicians and philosophers have supposed, or something else apart from this?"[66] One piece of evidence supporting the linkage of the pulse to breathing is that when a person exercises, both his breathing and pulse increase together.[67] Similarly, in rest, they decrease together. However, as we have seen earlier, Galen does not want to make the lungs the principal source of the body's *pneuma*. However, this link constitutes a way around this problem (since the lungs and heart are connected), viz., to make the *heart* the proximate actor in this scenario. Thus, for the most part, the etiology of pulse can be accounted for by the heart.[68]

However, the manner of this causation is simple. Since Galen wants to set out his own particular theory, and since the principal competitor is Erasistratus, much of Galen's account of the heart rests upon giving a unique interpretation. For Erasistratus, there were two separate vascular systems: the arteries for *pneuma* and the veins for blood (nourishment). The heart was also constructed so that it coincided with this. The left side of the heart was a *pneuma* pump sending the vital substance into the arteries, and this was the cause of the pulse.[69] Galen concurs that the blood vessels emanating from the left side of the heart were arteries and the right side were veins. (However, this is problematic with the designation of the "artery-like vein" for the pulmonary artery and the "vein-like artery for the pulmonary vein).[70] The reason for the problem is that the pulmonary vein is designated to be the *pneuma* carrier, which would make it an artery. However, its anatomical structure makes it a vein. Following theory over observation smacks of a *phusis*$_2$ cum D-□ account. Galen does not address this issue, but instead goes forward with the following explanation:

> The double and compounded motion of the arteries which we call the pulse is governed by the heart as we ourselves elsewhere (*De placitis Hippocratis et Platonis*, K V 4.3 & 4.7, cf. DeLacy, 1981), and many others before us have demonstrated; it is not so, however, in the manner supposed by Erasistratus, but as both Herophilus and Hippocrates, and nearly all the most esteemed of the ancient physicians and philosophers believed. For that power which is in the body of the heart by which it expands and contracts, flowing out through their coats to all the arteries expands and contracts them, just as it does the heart . . . There are many pores in the arteries, some in the coats themselves—vents, as it were—others, like mouths, open either into the guts or the stomach or into this outer skin. Moreover, the arteries are linked continuously with each other and / with the heart by means of very large pores, or rather,

> by means of all of their inner coats. To the veins, however, they are not linked through large pores. (Furley and Wilkie 1984, 210–211; K V, 163–165)

This passage *does* allow for the heart to be responsible for the *power* of the pulse, but not as a pump, but as a purveyor of that power through the layers of the artery (the endothelium, perhaps through the agency of natural *pneuma* from the heart?). The arterial system is said to be structurally unique with its layers and pores, which would also permit the whole-body skin breathing alluded to earlier. Thus, the anatomy of the blood vessel controls its activity and confers its unique function.

The story of how the pulse occurs links the heart to the arteries so that they are in unison: during systole the heart contracts, the aortic valve is open, and arteries constrict. During diastole the aortic valve is closed and the heart and arteries expand simultaneously. This is in contrast to Erasistratus who had the heart sending out *pneuma* pulses that followed consequentially from that, generating cardiac input.

During diastole the arteries draw in air, and during systole they expel *pneumatized* blood along with waste products—so-called muddy blood (generally venous).[71] This is not to suggest that the heart pumped blood, but that via the operation of the valves, it opened and shut the opportunity of blood to flow out via the directive powers of *pneuma* and attraction (by various bodily organs) to the rest of the body.

Galen's famous error in this regard is his conjecture that blood passed from the right to the left ventricle through the interventricular septum. The reason for this error might have been that Galen knew that the vena cava emptied into the right atrium and from there via the tricuspid valve into the right ventricle. *Some* of that blood would flow to the "arterial vein" (pulmonary artery). That blood was destined to both nourish the lung (with thin blood appropriate to the principal input of air) and to allow other blood to be enriched by air and to return to the heart via the "vein-like artery" (pulmonary vein). But what about the rest of the blood in the right ventricle? It needed somewhere to go. Following Erasistratus, Galen posited tiny tubes, *anastomoseis*, that existed all over the body. Here in the interventricular septum they also existed as a means of moving this excess blood. Galen is a bit unclear on the necessity, D-□, of this except to bring in the *phusis*$_2$ explanation that Nature does nothing in vain:

> The thinnest portion of the blood is drawn from the right ventricle into the left, owing to there being perforations in the septum between them: these can be seen for a great part [of their length]; they are like pits with wide mouths and they get constantly narrower; it is not possible, however, actually to observe their extreme terminations, owing both to the smallness of these and to the fact that when the animal is dead all the parts are chilled and shrunken. Here, too, however, our argument,

> starting from the principle that nothing is done by nature without an aim, discovers these anastomoses between the ventricles of the heart; for it could not be at random and by chance that there occurred pits ending thus in narrow terminations. (*On the Natural Faculties* III.15, Brock tr.)

This particular argument smacks of an a priori use of *phusis*$_2$ aspiring to the certainty of D-□. It is not conclusive and is a step backwards in Galen's espoused *phusis*$_1$ critical Empiricism, D-◊. However, what he does get right is the idea of pulmonary circulation with blood going to the lung and then returning to the heart. In other contexts he also had some keen insights into the flow of fetal blood movement in the heart. Galen correctly depicted that fetal blood moved from the vena cava via the foramen ovale to the left atrium.[72] He also notes that this movement ceases after birth.[73] Perhaps this fetal blood observation led him to conclude that blood could move right to left in the heart (atria and ventricles).[74]

Thus, the *how* of the pulse has been answered. It contains aspects of all three logical possibilities that were floating around at the time: (a) the agency of the heart, (b) the *dunamis* in the arteries themselves, and (c) natural *pneuma*. Galen hazards his own particular theory here based upon what others had said and his anatomical observations of the complex construction of the arteries. Because of their complex structure (unlike veins), they are accorded a more important role than was given by Erasistratus. As we assess the two today, it is certainly true that Erasistratus' account of the pulse is more in line with Harvey's reconstruction of the vascular system. But by giving the arteries a more central role in his account of the pulse, Galen is in tune with those modern scientists who think that the arteries are complex physiological entities that are too often marginalized as mere transport pipes.[75]

The answer to the *why* of the pulse relates to the body's need for nourishment and *pneuma*. One interesting question in this regard concerns psychic *pneuma* produced in the brain. Since *aer*$_1$ is a source material of psychic *pneuma*, and since psychic *pneuma* is extremely important for the operation of the body's functions, it is useful to understand where the brain gets its supply of materials for the creation of psychic *pneuma*. It was argued earlier that some of this comes from the nose and some from the heart in a process that was speculatively reconstructed. But Galen is keen to demonstrate that the heart source is certainly minor.

To show this, Galen performed a crucial experiment:

> [T]he brain does not need all the *pneuma* from the heart. It is accepted that either the vapor / rising to it from the blood is sufficient, or what is inhaled through the nostrils; but it is probable that the rising vapor too becomes scarce when the arteries are ligatured . . . so it must be that it is for the most part from breathing in through the nostrils that

> the nourishment [air_2][76]comes for psychic *pneuma*. . . For, covering the mouth and nostrils of a boy with a large ox-bladder, or any such vessel, so that he was unable to draw any breath at all outside it, we saw him / breathing unhindered through a whole day.[77]

In this experiment an ox bladder is placed loosely over a boy's head so that his normal breathing is impaired but not so tightly that he suffocates. It is just tight enough so that there is a marked difference in the amount of fresh air coming to the boy. The point seems to be that if the production of psychic *pneuma* is dependent upon fresh air, then there should be a marked difference in the boy's mental state when he breathes stale air. However, Galen reports this not to be the case. The boy continues normally throughout the day.[78]

Everything here depends upon how tightly the bladder is positioned over the child's head. The closest modern analogue to this would be having a person who is hyperventilating breathe into a paper bag covering the nose and mouth.[79] This is generally administered for up to a minute (clearly not for an extended period of time). Overextending the time would lead to a situation in which the child would be breathing carbon dioxide–rich air. This cannot go on for too long without there being signs of mental disorientation. However, if the bladder were not very tight, then the boy might continue naturally, but then the point of the experiment is lost (i.e., the supply of outside fresh air is not diminished).

This is indeed a troublesome experiment.[80] It seems that either this is an experiment that Galen did not really perform (a sort of thought experiment) or it is one in which a particular outcome was so desired that the methods and materials of the experiment were manipulated so that the hypothesis was confirmed. In either case, this use of experiment by Galen is not conclusive.

One might conclude that though the careful description of the pulse by Galen created a valuable means for diagnosis[81] that has only just recently been abandoned by medicine, many issues in the philosophy of science concerning the *how* and *why* of the pulse do not constitute his finest work. This may be because he was overly concerned in making his own reputation (as superseding Erasistratus) and because he offered experimental evidence that is rather suspect on several counts. What is clear, however, is that Galen affords the arteries a power of the pulse and thus endows the arteries with a complicated set of functions: carrying blood, carrying *pneuma*, and engaging in the pulse. These, together, suggest a unique vascular complexity that certainly elevates the physiological importance of the arteries.

C. Arteries

The third question is why the arteries are not merely pneumatic tubes. The arteries have been and are presently (in the twenty-first century) a subject of some controversy. Part of the problem lies in the anatomy of arteries with their various layers. The structure of the arteries is much more complex than the veins. For example, the endothelium is very independent in its character and is seen by some modern scientists as perhaps being a biological system by itself.[82] Whether it is or is not must be the topic of another discussion.

When Galen took up this question, he was keen on disputing Erasistratus' account. Erasistratus set out a rather well-thought-out theory of how to get *pneuma* to the body and that the arteries carried only *pneuma*, while the veins transported the blood to the tissues to be used up as nourishment (see Chapter 4, this volume).

How does one confront a theory that asserts that the arteries are (for the most part)[83] pneumatic tubes? Two experiments that Galen describes may give an indication about his method in solving the question.[84]

In the first experiment described earlier in Chapter 4 in this volume, Galen calls for tying the artery in two places and then inserting a narrow tube into the artery.[85] If the artery were only filled with *pneuma*, then one might expect that he or she might witness a sound of air escaping. Instead, what one sees is blood escaping. In one way, this might suggest that the artery carries blood along with *pneuma*.

On the other hand, Erasistratus had (perhaps?) done (or thought about doing) this experiment, too. His account was to say that because nature abhors a vacuum (a large explanatory principle), making a hole in the artery creates a situation in which the blood in the adjacent vein would move into the artery through its lesion. Thus, according to Erasistratus, if one causes trauma to an artery, it is perfectly understandable that what is observed is blood and not *pneuma* escaping. What we have is a theory that is not falsifiable. Though it is far from a perfect evaluation, falsifiability is generally recognized in modern times as one test of a scientific theory.[86] According to Erasistratus, all the observed effects of the first account would be the same. However, there is a different *reason why*. Galen is seeking an empirical test of the pneumatic tube hypothesis.

To do this, Galen tries to counter Erasistratus' objections by suggesting that Erasistratus' experiment mentioned in Chapter 4 in this volume (where one ties an artery, makes an incision, and then inserts a small tube) does not illustrate Erasistratus' point that the arteries are only filled with *pneuma*. This is because by varying the size of the incision, the results remain constant, which belies the vacuum reply of Erasistratus.[87] Galen's refutation of the experiment in Chapter 4 in this volume focused upon the procedure of the experiment. But he also extends his attack to analogous cases involving trauma. If the hole were very small, then the trauma caused would be small and it would be unlikely that blood would *immediately* flow unless the arteries carried blood. In the second case Galen observes what happens when a major or moderate wound occurs where the blood flows freely. Galen observed that there was a vapor associated with this, but not the sort of evacuation predicted by Erasistratus' theory.[88] The vapor or fine mist that seeps gently out, Galen contended, must be *pneuma*.[89] This would demonstrate that *pneuma* coexists well with blood. In each case (the small incision and the gross trauma), the predicted result of purely pneumatic arteries was contradicted by the results of the experiment.

What is especially problematic with the method of this experiment is the single perturbation experimental model. Since any physiological event has a number of dependent and independent variables, the single perturbation account ignores these and is thus ripe for error and multiple, inconsistent interpretations.

In the case of these experiments, Galen is calling for a reinterpretation of the doctrine of Praxagoras and Erasistratus that the arteries carried only *pneuma*. On the other side of the question stood Herophilus who claimed that both the arteries and veins contained *trophe* (nourishment) conveyed by blood.[90] The history of science often rests upon these sorts of moments: (a) should one accept the existing theory—here being Erasistratus (even modified greatly with ad hoc side modifications), or (b) should one accept the alternative theory that can account for the data in a different manner without ad hoc modifications?[91]

So how does one decide? First, in Galen's account, there is a violation of the dictum that each part has only one function (i.e., the arteries are conveyors of both *pneuma* and blood). This would seem to tell for Erasistratus. Second, there is a violation of the dictum that Nature does nothing in vain. If redundancy means unnecessary complication, then simplicity would argue against two sorts of vessels carrying blood. This argument scores again for Erasistratus. Third, the analogous experiments (the small incision and major trauma) seem to more plausibly falsify Praxagoras/Erasistratus (see Chapter 4, this volume). Score one for Galen.

Thus, what we are left with is this: (a) should we rest on perfectly sound reasoning from accepted theory on the principle that the observations may be incorrectly interpreted or that there was a flaw in the procedure (the D-□ modality)? Or (b) should we trust our experience (after repeating our tests and checking our methods as much as possible—E-◊)? As this book argued in Chapter 4, this was a major problem in the ancient world and (along with some version of skepticism, Table 4.1) gave rise to the major schools of Dogmatism, Empiricism, and Methodism.[92] It is also a recurrent theme in the history of science from Copernicus to Newton to Darwin to Einstein to modern particle physics and genetic engineering.[93] It is by no means an easy call. Sometimes what appears to be clear empirical evidence proves to be false due to theoretical failures.[94] On other occasions we have seen those clinging to theories in spite of falsifying evidence.[95]

In this instance, the theoretical points may sway to the established position of Erasistratus, but in the end, the systematic challenger, Galen's, falsification spin on the experiment wins the day for Dogmatic reasons: it is more primary to the *explanandum*.[96] Since most modern criteria for epistemological justification in scientific reasoning supports Galen's general move here, his argument wins modern approval.

The final point of exploration in this subsection is the medical treatment of bloodletting. Bloodletting began with the Hippocratic physicians, but widespread practice took some time.[97] The condition is called *plethos* or *plethora*, that is, "abundance," or more accurately in this context, "over-abundance." When there is too much of any humor, then a remedy in the opposite direction is needed.[98]

A clinical example of how and when venesection might be called for comes from *De methodo medendi* IX, 4. A young man who is very regular in his exercise began experiencing fevers and insomnia. His pulse was regular, very large, rapid, frequent, and strong. It seems that the young man had failed to exercise for the previous month and then exercised vigorously—though not to excess. His diet had remained unchanged, but his digestion was not proceeding properly. Because his pulse was regular, his fever constant (and not increasing), and his urine clear, Galen decided that the young man was not yet suffering from putrefaction (*sepsis*). In order to ward against this possibility, Galen decided to engage in venesection:

> Accordingly I deliberately took from him an amount sufficient to bring about fainting, since I had been taught that both on theoretical grounds and from experience this is a most important remedy for continuous fevers in someone of strong capacity. For first, a body being cooled in the fainting very quickly comes to the opposite state. Nobody would be able to discover anything more pleasant or useful than this . . . those present said, "You, sir, have slain the fever!" (K X, 612, Johnston and Horsley, tr.)

There are several important points to bring forth from this short passage. First is that the venesection is carried out for the purpose of causing the patient to pass out. This is a procedural goal that is dangerous because there is quick transition when one loses blood to brain damage and even death.[99] Galen was well aware of these adverse outcomes.[100] This is why the physician needs more than theoretical knowledge, D-□. Actual evidence-based medicine is needed, E-◊. This shows that Galen, the synthesizer, had to bridge these two categories in his medical practice—leaving him at D-◊.

So now we have a physician who knows just how much blood to drain (according to the soma type of the patient). The blood is drained from a vein, and the patient faints. Now this experience of fainting has the following mechanical effect. First, the patient becomes colder because there is less blood (a source of heat) within his body. This is important because the patient is being treated for a fever. Now it is also important that the patient is someone of strong capacity. One would not propose such a treatment on someone of greatly advanced age who was feeble. The shock to the body would create a greater harm than the "reset" brought about by the patient's fainting (becoming cooler, and then beginning again). For Galen, this mechanical account follows the explanation-type α account in the mode of *phusis*$_1$ (see Chapter 2), this volume.

As bloodletting was a popular remedy among other practitioners for some time, they would not be able to give the same sort of mechanical account (characteristic of Dogmatic practitioners). Instead, they would be forced to accept explanation-type β and its reliance upon *phusis*$_3$.

Galen wrote three works devoted to venesection: *De venae sectione adversus Erasistratum* (during his first time in Rome), *De venae sectione adversus Erasistrateos Romae Degentes* (during his second period in Rome), and *De curandi ratione per Venae Sectionem* (a very late work).[101] In the first work Galen's principal aim is to rail against Erasistratus—and especially his followers in Rome for not practicing venesection. The second work is somewhat less polemic, while the third is rather businesslike.[102]

The rationale for venesection as a medical treatment begins with the blood in the veins because that is where the muddy blood (blood filled with *perittoma*) resides. Muddy blood can lead to putrefaction, *sepsis* (a cause of various medical disorders—principally fevers). Arterial blood is vibrant with *trophe* and *pneuma* and so it would not make sense to relieve someone of that blood. Now as we have seen in Chapters 2 and 4 in this volume, putrefaction was cited as a major cause of disease from the Hippocratic Writers onwards. This toxic residue causes blockages, infects various regions of the body, and otherwise disrupts the somatic balance. What are the causes of excess *perittoma*? Well, residue is that which is not used or turns away from its useful purpose (see Chapter 3, this volume). If one restricts the amount of food eaten, then there will be less residue as a result. Thus, one way to counteract the residue in the venal blood (the slow evacuation strategy) is to fast. However, this can take time. If the patient is faced with a critical condition, then fasting might not kick in soon enough.[103] The patient will die.[104] This is essentially Galen's argument against Erasistratus in the first of his bloodletting texts. The putrefaction within the venal blood must be eliminated when the patient is very hot or suffering from inflammation. An evacuation strategy that is efficient is to be preferred to one that is slow and indirect. Bloodletting is the efficient evacuation strategy.[105]

Of the sort of fevers that a patient experiences, many are superficial and only require topical cooling such as wet compresses.[106] It is the inflammatory fevers that are more dangerous. Bloodletting allows the physician to move the focus of the putrefaction that is causing the fever.[107] The theoretical point behind this at the vantage point of the venal system is to create a novel flow that dislodges the blockage. Bloodletting does just that—which is *why* the patient becomes cooler and experiences a *reset*. Blockage of the normal body flow of blood was thought to be a major cause of disease.

Other common instances of blockages that required their own nonblood intervention strategies were constipation and late menstrual flows. Since remedies for these conditions were rather difficult, patients who tended to have these sorts of blockages were encouraged by diet and exercise to create a preventative prophylaxis.[108]

A final indication for bloodletting is for *redirection* or *revulsion*. This was particularly important when treating a gladiator or soldier who was severely wounded. The patient is bleeding to death due to an involuntary trauma. If the physician opens a vein in another part of the body, then the bleeding from the primary wound will abate. The new site of bleeding (the opened vein) can be stopped anytime by the physician. Thus, an uncontrollable fatal situation may become manageable.[109] Galen had experience working with gladiators and soldiers. However, this treatment does not work as promised.[110]

The whole enterprise of venesection was widely practiced both in the West and in the East for centuries.[111] It follows logically from humor theory supplemented by the doctrine of *perittoma* and its resulting *sepsis*. The small amount of blood loss that is consequent to bloodletting will generally not hurt a patient; however, it does not help either except via placebo effect (bolstered by the patient taking greater care of himself because of the traumatic event of fever or inflammation). This is certainly a case in which a significant intervention is set forth in which all parties hope for success. Though venesection was widely practiced by D-□ and E-◊ practitioners from the explanatory-type β perspective, it is the synthetic, systematic Galen who weighs in as D-◊ under the rubric of explanatory-type α.

CONCLUSION

This chapter has sought to explore philosophically some key points concerning Galen's arterial physiology. What Galen does right is that he often utilizes a synthetic and systematic explanatory scheme to account for natural phenomena following for the most part a model of critical Empiricism (D-◊ and explanatory-type α). However, he is not as rigorous in this regard as Aristotle. This may be because Galen sought to stake his own claim to medical greatness respecting blood and its environments by vanquishing his rival Erasistratus (and his numerous followers). Galen saw no real competition from Plato, Aristotle, Hippocrates, Praxagoras, or Herophilous, et al. Erasistratus was the target. Sometimes, in the case of the pulse, this put Galen into some difficulties. However, his synthetic presentation method of bringing in supporting material from his predecessors, *historia*, along with his own particular twists based upon his own extensive observations of patients and gladiators[112] (with their injuries), allow him to offer more than a handbook of what *others* have put forth. Indeed, it is the mark of a systematic thinker first to take into account what others have said before him and then to offer key distinctions that will satisfy long-standing debates through consistent posits that exhibit wide explanatory power.

What Galen did do in this instance is to show that the arteries carry blood and *pneuma*. He reaffirmed and expanded upon the complicated anatomical structure of the heart and arteries. The anatomical structure of the arteries allowed for a complicated transfer of *dunamis* from the heart to the arteries so that the pulse is possible as an aid to the transport and a fortiori offers the physician an important diagnostic tool.

Galen also offered a rich theory of respiration that included skin breathing and direct nostril air transport to the brain (to supplement normal lung breathing). Thus, for a theoretical system that favored one function for one part, Galen presents a rich variety of functions for the arteries. This amounts to a significant elevation of the arteries such that one might say (without exaggeration) that though there are three principal organs in the body—the brain, the heart, and the liver—the arteries stand by as a strong supporting actor.

Galen saw blood in its various forms: from muddy to nutritive to rare and full of *pneuma*. Each formed a critical part of his entire medical explanatory apparatus. It was, after all, the most important humor. The extent that his writings were recopied over the ages (at great cost) yields proof of how powerful and successful his synthetic and systematic approach has proven to be. Galen's work stands as the terminus of biomedical practice in the ancient Western world, and the principles it embodies still challenge contemporary categories within the philosophy of biomedicine.

NOTES

1. For example, in *The Natural Faculties* I. 16 Galen attacks Erasistratus at great length. He does this to set out his own position. However, he does praise Heraclides of Tarentum (an Empiricist) and Soranus (a Methodist) even though they represent positions with which he disagrees; see Nutton (2013): 151, 200.
2. It should be noted here that there is some controversy concerning Galen's method of exegesis and whether he was consistent and fair in his commentaries. For a careful discussion, see Rebecca Flemming (2008). Another point that is often made is Galen's interest in situating himself among developments in philosophy; see G. E. R. Lloyd (2008).
3. For a refresher on these terms, see the glossary with references to earlier chapters.
4. One key exception to this is the *Theaetetus*. This is one of Plato's key dialogues on truth theory. He situates real people and actual positions that are being discussed. For more on this, see Jenkins (2011) and Nawer (2013).
5. In Western philosophy three of the most prominent systematic philosophers are Aristotle, Aquinas, and Kant. In each case, the methodology of the writer is to summarize what had come *before* and critically evaluate it—the cons and the pros. The good is adapted to the general program, while the bad is discarded. The result is an original, synthetic unity.
6. Ryle: 35, cf. Robjant (2012).
7. It is not necessary for the nonsynthetic but systematic philosopher to totally *ignore* everyone who came before, but it is the case that they do not think that the synthetic project is an essential propaedeutic to the systematic project. Galileo, for example, sets out his vision and then pokes at those who disagree—such as his version of Aristotle. It should also be clear that under this conception of philosophy of science that there are no prominent philosophers who are synthetic but nonsystematic. The eristic philosophers are nonsynthetic and nonsystematic.
8. I am thinking here of the intellectual climate, but others such as Van der Eijk (2005a): 1–44, would extend the context to cultural and literary themes

following an inclusive post-modernist agenda. While this scope may yield some interesting results on the broader synthetic context, it is beyond the context of this chapter.

9. *Republic* 4 .435–42, 9. 589d, cf. *Laws* 9.863b, *Republic* 6.504a, 8.550b, 9.571b. Plato puts the soul into the body at *Timaeus* 69e ff. For a discussion of this, see Hankinson (1993a, 1991).
10. K 11, 379; 12, 377; 12, 378–13, 361, 362–1058; see also Nutton (2013): 250–253.
11. I should remind the reader that Aristotle himself saw the brain as having a more limited function of regulating blood temperature, PA 652b 22.
12. On the face of it, there seems to be a problem in asserting that the soul is tripartite with organ locations and the doctrine that the soul is one, incorporeal, and indivisible. For a discussion of this, see Tieleman (1998) and Gill (1977).
13. This is one of the most important explanatory concepts in Aristotle. Galen takes it on wholeheartedly. The reader should note that the capital 'N' for nature refers to $phusis_2$ when talking about Nature as a creative biological force and a lowercase 'n' when discussing the properties that some organ or individual organism possesses. For a further discussion of this, see Boylan (1983, 69, 87, 105); see also the discussion of $phusis_2$ in Chapter 1 of this volume.
14. One must be careful with the word 'mechanical' here because in Galen's context a mechanism might have a vital spirit as part of its material account. This is certainly different from the way the term has been used from the seventeenth century onward. For a recent discussion of some aspects of material explanation in Galen, especially in the interplay with teleology, see Berryman (2002) and Cosans (1998). For a defense of the concept of "mechanical" in the ancient Greek world, see Berryman (2009).
15. Simian anatomy is very similar to human anatomy—but not on all accounts. The most interesting difference is in the circle of Willis in the brain. However, since this chapter is not about proving which particular points of anatomy or physiology that Galen got "right," but rather is about the general notion of the arteries, blood, and pulse, this note should merely be regarded as a caveat. Note also that when Galen studied in Alexandria he says that he also engaged in human dissection. However, this was a place and a moment in time that was rare in its tolerance of such procedures; see Chapter 4 in this volume.
16. It should be noted that this is one large direction in which Galen diverts from Aristotle, that is, that the brain and not the heart (Aristotle) is responsible for cognitive thought.
17. *De placitis Hippocratis et Platonis* K V, book 1 has the most extended discussion (see also DeLacy, 1981), cf. *De anat. Admin.*, K. II 667–698; *De motu musculorum* K IV 367–464.
18. One counter example to this is the vena cava vein that goes from the heart to the lung. Variously, this is described as an "artery-like vein" (*De use partium* VI, I. 308–311) or else having a very special function of bringing blood to the lungs. However, in *De naturalibus facultatibus* I.77 (K II), a "vein-like" artery seems to be the pulmonary artery. Obviously, there is a problem with technical terminology, but this shows the tension between observed anatomy and preconceived physiological function.
19. *De usu partium* May 89, (H I, 33); 205 (H I, 197); 207 (H I, 200); 220 (H I 217), cf. *De plac. Hipp. Et Plat.*,VI.5 (K V 568ff); see also DeLacy (1981) and Nikolaus Mani (1959).
20. Sometimes Aristotle is ambiguous with reference to the material cause, as in *Posterior Analytics* 2.11.

21. *Physics* 194b 16–195b 30, cf. *Metaphysics* 983a 24–32; *Posterior Analytics* 94a 20-b 27.
22. Boylan (1983): ch. 3.
23. PA 639b 14, 642a 2, GA 715a 9.
24. *De symptomatum differentiis* (K VII, 47–48) and *De sectis ingredientibus* (K I, 64)).
25. *De usu partium* (K III, H I 340); see also Darenberg (1854 I 420–422) on the causal nature of Galen's addition cf. *De causis procatarcticis* vi, 59–70 and Hankinson (1998 91–95). It may be impossible to obtain a high level of precision on this because of the complex contexts in which the terms are used—see Striker (1986).
26. cf. Hankinson 1998c.
27. Boylan (1983).
28. Christopher Gill (1983) and (2009), cf. Harris' account of the Pneumatist school (who were influenced by the Stoics)—Harris (1973): 235–251.
29. K VII, 826, cf. *De usu respiraionis* 1. It should be noted that the word *diapnoe* may merely refer to the disbursement within the body as opposed to the dispersal in the acquisition mechanism.
30. *De difficultate respirationis* II, K VII 826, *De usu respirationis* 5.4, K IV 506 (see the edition by Furley and Wilkie), cf. *Peri phuson* 12, line 1, the Hippocratic Writer describes breaths passing through the flesh.
31. K IV 506, cf. *De usu pulsuum* K V 166, Hippocrates, *Epidemica* 6.6 (Littré V 322).
32. David Furley suggests that, at best, the Hippocratic basis for Galen's doctrine is rather shaky (Furley and Wilkie 1984: 12).
33. For an overview of some of this controversy, see Smith (1990). Geoffrey Lloyd has mentioned to me that my light skepticism in this regard can really lead to some crucial misleading understandings of historical influences concerning whether there is really any conflict between Coans vs. Cnidians. This is certainly a disputed issue.
34. For a support of my reading of Galen as both a synthesizer and a Dogmatist, see Teun Tielman (2008).
35. It should be noted that Galen also allowed for some $phusis_2$ accounts—particularly when discussing conception theory (*Foet.Form.* VI 693), cf. Boylan (1986a), Flemming (2007). Galen called his final treatise on teleological anatomy, *On the Usefulness of Parts*, a poem to the Creator of Nature, K IV, 365–366.
36. *De placitis Hippocratis et Platonis* VIII 8 (K V 708); see also DeLacy (1981).
37. Berryman has suggested to me that my insistence upon a material account here is unfair to Galen. He may be satisfied with a nonreductionistic account. However that may be, my depiction of Galen as adopting an explanation-type α account would commit him to the broadest set of material causes possible. However, it is possible that at times Galen adopts a $phusis_3$ approach and thus would not need additional rationale beyond the *holke*.
38. This process may not have been as smooth as I suggest. Nutton (2005) argues that the humor theory of the Hippocratic Writers was modified by Galen and then modified continually in Byzantium, the Muslim world, and the later Latin Middle Ages, cf. Schöner 1964; Kiblansky, et al. 1964. It is also possible that the *serum* option connects to the discussion of *ichor* from Chapters 1 and 2 in this volume.
39. It should be noted that this citrus response was not supported by Galen. Galen's father instilled a bias against fruit (due to the father getting sick on unripe fruit on several occasions). See Nutton (2005).

40. Both Nutton and Lloyd have noted to me that we must be very careful here in suggesting that there is a "set humor theory." I agree. However, when we segue to philosophically analyzing various theories in the philosophy of science, it is sometimes useful to posit a position to logically examine even if it is really an examination of certain existent texts and setting out various coherent interpretations within certain categories. So long as we all agree that many exceptions exist, I still believe that this is a useful exercise.
41. Compare to the discussion of clinical medicine in Chapter 2 and of skepticism in Chapter 4 of this volume.
42. When the disease is primary, it is the Methodist approach; when symptoms are primary, it is the Empirical approach. These at times can overlap—see Chapter 4, this volume.
43. There is a key equivocation with *phusis* as per Chapter 1 of this volume. There, three key senses are set forth in epistemological terms that imply ontological realities. In this instance, the *type-token* issue is to the fore (see Chapter 2, this volume).
44. The definition of "disease" and "health" are rather complicated concepts; see Boylan (2014a).
45. Of course, the Skeptics in the ancient world would demur (see Chapter 4, this volume) and David Hume (tr. 1977): 50–53; in the modern world, they do the same.
46. The Pneumatists (influenced by the Stoics) are also a philosophical force, but their influence in the philosophy of science, according to this author, is rather derivative of aspects of the Dogmatic Hippocratic Writers, Aristotle, and Diocles. For a contrary view, see Harris: 235–251.
47. Boylan (1986a): 44–77; Van der Eijk (2005a): 28–29, 299ff; Cosans (1997): 35–54.
48. Diels and Kranz I. 282, 286, cf. Aristotle, *De generatione et corruptione*, 328b 26–338b 19, *De partibus animalium*, 646a 12–24; Galen, *De naturalibus facultatibus*. I. 2. 3 (K II 4–5, 7–9).
49. Boylan (1982).
50. Boylan (1982).
51. The way the blood moves for Galen is largely a feature of attraction. However, the fact that Galen stipulates that the heart and arteries dilate and contract at the same time has given some impetus to setting Galen as a sort of missing link between Aristotle and the Anatomists and Renaissance work that led to Harvey. For a few contemporary accounts, see Wright (2013), Aziz et al. (2008), Limet (2010), and Aird (2011).
52. Furley and Wilkie: 4–6.
53. I remind the reader (as per the preface) that these modern terms of analysis are not meant to represent the worldview of ancient practitioners of medicine. Instead, it is a philosophical device to timelessly evaluate ancient scientific texts.
54. As was suggested earlier, there is also air that enters the arteries via pores in the skin that are transported to the thin arteries around this area. From there, this air (under my rather speculative account) undergoes a change to aer_2 that is equivalent to $pneuma_1$. And then the entire process is repeated.
55. The use of this word is to make an unsubstantiated yet speculative connection to the pluripotential stem cells of present-day biomedical research. If I am correct in my conjecture about the transformation of *pneuma*, then at this first stage there is a chemical change that opens the air, thus transformed, to be able to become hepatic *pneuma*/natural *pneuma* or psychic *pneuma*/natural *pneuma*. This can be similar to our present disposition to stem cells in biomedical research. For an introduction to this, see Boylan and Brown (2002, ch. 10). Readers should note again the cautions of the preface and fn 53.

56. This would include present neurological functions and a *spirit of life* understood as a hybrid of nonmaterial soul and some material delivery vector. The transfer and delivery mechanisms are left as mysterious, *Passions of the Soul* I.10–11; 31–32; cf. to Plato's less famous conjunction of the two realms in the bone marrow, *Timaeus* 73b.
57. Nutton (2013): 239–240.
58. *On Temperaments* 1, 509–694, tr. P. N. Singer (1997): 202–289.
59. *Dunamis* is generally understood in Aristotle to mean 'potentiality' and is linked with *energia* and *entelechia* that refer to two types of 'actuality.' In Galen, the term has a somewhat different meaning of 'faculty.' A faculty (as per Brock 1916) refers to the ordinary activity of a part. This is rather different from Aristotle, who understood potentiality over a broader time span, for example, an acorn is potentially an oak tree. For Galen, the example would be "the heart via *dunamis* causes the pulse to occur in the arteries." This second sense implies a regular power that is about to be expressed.
60. Here I am following D. Martini (1962, VIII, 41–47, 41).
61. *De placitis Hippocratis et Platonis* K V 505–585 (see also DeLacy 1981), cf. Plato, *Republic* 435–442 and *Timaeus* 69–72.
62. *De usu partium*, VI.17 (May, p. 324).
63. Here and other places it is often unclear whether Galen is setting out his position because it is entrenched within his theoretical structure or because he wanted his own unique position.
64. *De usu partium*, K III, 4.9.
65. *De usu partium*, K. III: 445.
66. Furley and Wilkie (1984, 195; K V, 149).
67. Furley and Wilkie (1984, 197; K V, 152).
68. However, there are exceptions. For example, in *De pulsibus ad tirones* K VIII, 474, Galen talks about the effect of bile upon the pulse. This would mean that the liver (the source of bile) also affected pulse. (Or at least could do so in certain circumstances such as anger.)
69. Harris (1973, 232). Harvey concurred with Erasistratus on the origins of the pulse (1957, see *Prooemium*). It should also be noted that the right side of the heart (right ventricle) did not act in exactly the same way for blood. But in this instance we are concerned with the pulse and arteries.
70. Harris (1973): 281, suggests a logical fix to this if the pulmonary blood vessels could be distinguished in the lung rather than their connection to the heart—which approximates the strategy of moderns of the "coming toward" vs. "going away" distinction. However, these actions assume circulation of blood, which Galen did not put forth. Galen, the synthetic thinker, uncritically accepts Herophilus' criterion.
71. For a longer account, see Harris (1973): 332–333.
72. *De usu partium* VI. 21.
73. *De usu partium* XV, 6.
74. For a more complete discussion of fetal blood flow, see Harris (1973): 287–295.
75. For an account of the important role of the endothelium, see Hwa, Sebastian, and Aird (2005).
76. My brackets.
77. *De usu respirationis* (K IV 504–505) aka *De utilitate respirationis* (Furley and Wilkie, 1984, 124–25). Galen also offers as evidence for this the tying off of the carotid artery leading to the brain—supposedly showing the same thing, viz., that the *pneuma* from the heart (via the lungs and skin pore breathing) is not really necessary as a source for psychic *pneuma, ibid.* K IV 502–503. For a slightly different reading of this experiment, see Cosans (1998 63–80). Cf. also to Erasistratus' other experiment discussed in Chapter 4 of this volume.

78. This experiment has obviously caused controversy. The Royal Academy worked at this to see whether certain controls might change the outcome—such as controlling the amount of air exchange. See Birch (1756 I. 179).
79. I am told by William Mullins, MD, that this is an archaic treatment and not much used presently. Perhaps this treatment owed its genesis to Galen's experiment?
80. Furley and Wilkie (1984, 256–257) comment on this passage and suggest that it may be spurious. Noll (1919, 1246–47) suggests that the bladder is meant to cover up the pore breathing of the skin, but this misses the point of the passage that seems to focus upon nostril as opposed to lung breathing.
81. *De symptomatum causis* K VII, 85–272; *De tremore, palpitatione, convulsione et rigore* K VII, 584–642.
82. Hwa, Sebastian, and Aird (2005).
83. Even Erasistratus admitted that blood might enter the arteries, but he thought that *according to nature*, this was not the case (*An in arteriis* K 712).
84. One is never exactly sure in ancient biomedicine whether the writer actually performed the experiment he said he did. It is very possible that it is merely a thought experiment. We must keep this possibility in mind.
85. *An in arteriis*, K IV, 713.
86. The literature here is hopelessly large. Classic examples include Popper (1959[1934], I.6); Carnap (1950 sect. 1–6); Hempel (1945 1–26, 97–121); and Goodman (1955 chs. 3–4).
87. (*An in arteriis* 2, 3, 8, K IV, 706–709, 733; *De anatomicis administrationibus* 7.16, K II, 646ff.
88. K IV 724.
89. Of course, today we would associate the vapor of a major laceration with the differential in temperature of a warm liquid entering a cooler environment and thus causing condensation. (Though because this is the Mediterranean climate, the differential wouldn't be as great as those replicating the experiment in Sweden during winter, for example.)
90. Anonymus Lodinensis (Jones 1947, 28.46–29.34). I am assuming that, following Aristotle, Herophilus intends *trophe* to be the same as blood.
91. The role of ad hoc modifications in the history and philosophy of science has been much written about. Some, like Hilary Putnam have declared that is more rational to set out ad hoc side hypotheses than to junk existing theory. Putnam (1974, 221–40) argues against Popper's thesis in the *Logic of Scientific Discovery* by discussing the actual behavior of scientists when Uranus' orbit was found to be incorrect (before the discovery of Neptune). This behavior was directed at modifying existing theories rather than discarding them. Thus, the goal of "all or nothing" does not describe the way scientists act.
92. From the groundbreaking work of Edelstein (1967) on this, there is an ongoing discussion. See Frede (1986) and Hankinson (1992).
93. Boylan and Brown (2002).
94. One famous case of this in the history of science can be found in the "discovery" by René-Prosper Blondlot in 1903 at the University of Nancy in France. Blondlot thought he discovered a new sort of radiation that affected the brightness of the spark in an electric spark gap. He named this new sort of radiation "n-rays" and received a prestigious prize as a result. It wasn't until a decade later that a US scientist, Robert W. Wood, showed that "n-rays" didn't really exist. A theoretical mistake affected the measurement devices and created the appearance of rays that didn't exist. For a further discussion of this, see Gardner (1957) and Ashmore (1993).

95. Helen Longino (2002, ch. 2) makes the case against Karl Popper's famous falsification account.
96. Of course, if one were dead set against D-□ and D-◊ reasoning, then the response would be skepticism$_3$.
97. Brain (1986): ch. 6.
98. An overabundance of yellow bile, black bile, or phlegm created *cacochymia*, which also had to be treated by reference to opposites—generally via diet.
99. Humans differ in the amount of blood they carry. A typical average is between 4.7 and 5.0 liters. The volume is largely controlled by body mass, and within individuals of the same body mass, the kidneys are responsible as they respond to changes in renal arterial pressure (Hudnall 2012). Most people pass out after losing 20% of their blood. Brain damage can occur with a 30% loss, and a 50% loss often leads to death, *ceteris paribus*. For a detailed discussion suitable for students, see Hoffman, et al. (2012).
100. For the disability associated with brain damage, see *Meth. Med.* IX, 10—K X, 36368. Regarding patients' deaths, see *Meth. Med.* IX, 10—K. X, 367–68; *Car. Ven. Sect.* 12—K XI, 288–289.
101. Ilberg (1897) puts it after *De methodo mendendi.*
102. Brain (1986): ch. 5.
103. K XI, 156.
104. K XI, 163.
105. K XI, 156.
106. K X, 665–667, K VII, 296–297.
107. K X, 661–63.
108. These conditions continue to be treated in this same manner today.
109. Marganne (1980): 115–130.
110. In conversation with Arianne Boylan, MD, a neurotrauma surgeon, this treatment will not have the desired effect and will only result in killing the trauma victim more quickly.
111. For a popular account of bloodletting in the West, see Dobson (2013). For some context of venesection in China, see Kuriyama (1995, 2002) and Hong (2014).
112. Mattern (2013): 81–97.

Glossary of Philosophical Terms

Aristotelian causation (see Chapter 3). Aristotle offers four modes of causation: material, efficient, formal, and final. In Aristotle's biological writings, these are combined to form two accounts: ME (mechanical/material explanation) and TE (teleological explanation). Because in his practice Aristotle combines ME and TE, his balanced position is D-◊, critical Empiricism.

ME is the combination of the material and efficient causes. It highlights explanation-type α, the ontology of *phusis*$_1$; the epistemology is the careful scrutiny maxim, is token/type balanced, and aspires to D-◊.

TE is the combination of the formal and final causes. It highlights explanation-type β, the ontology of *phusis*$_2$; the epistemology is the selected faith maxim, is type-oriented, and aspires to D-□.

Biomedical theory/practice (see Chapter 2)

Clinical approach: The clinical approach emphasizes the practitioner's own evidence-based medicine supplemented by the clinical writings of others. This sets this modality in realm of contingency, ◊. It is aligned to the Empiricists. It is token oriented, embraces an extreme form of the careful scrutiny maxim (epistemology), and ontologically it is aligned with *phusis*$_3$.

Philosophical approach: The philosophical approach emphasizes the creation of theories to explain the unseen. Epistemologically it fits into the D (Dogmatist/ Logical) approach rather than the E (Empiricist) approach. When the philosophical approach is oriented toward the ontology of *phusis*$_2$, it seeks the modality of necessity, □ (which is type driven). When it is ontologically oriented toward *phusis*$_1$, then it veers toward contingency, ◊ (which is balanced between type and token).

Epistemology (see Chapter 4): The two main camps are the careful scrutiny maxim advocates and the selected faith maxim advocates. These describe the strategy of the source of knowledge.

Careful scrutiny maxim: An epistemological stance that seeks the most modal certainty as possible (D-□ is highest, D-◊ is next, and E-◊ is last). However, these proponents are very skeptical about the possibility of D-□. Therefore, this epistemological position leads to skepticism$_2$ and skepticism$_3$.

Selected faith maxim: An epistemological stance that strives to gain as much truth as possible. Proponents adopt explanatory-type β and seek the ontology of *phusis*$_2$. They are type oriented and are inclined to the modality of D-□. They are thus inclined to skepticism$_1$.

The forms of biomedical explanation (see Chapter 2)

Explanation-type α: A sequential material/mechanical causal process. It assumes the ontology of *phusis*$_1$ and fits into the modal continuum at D-◊.

Explanation-type β: An emergent process that embraces the ontology of *phusis*$_2$ (when the *explanans* is otherworldly or based upon magic) or *phusis*$_3$ (when the *explanans* is constrained by skepticism$_3$).

Modality of the scientific enterprise (see Chapter 2): There are two critical factors that set out the points on the modality continuum. First is the strategy of the *explanans*: either logically based—here represented by "D" indicating the Dogmatists/logically-based practitioners, or it is Empiricist directed—here represented by "E" indicating the Empiricist practitioners. Second is the supposed logical force of the claims from "necessary" (indicated by the □ quantifier) to some level of "contingency" (indicated by the ◊ quantifier). These two factors are combined on a continuum of modality.

D-□: A deductive-nomological position of certainty based primarily upon logical reasoning.

D-◊: A position of critical Empiricism. It seeks to embrace a logically based systematic account (the "D") along with a commitment towards empirical investigation, for example, dissections, case studies of disease progressions that shows an *epi to polu* (for the most part) attitude (the "◊).

E-◊: A clinical medical-practice position based upon what has worked in one's own experience as well as the *historia* input of what has worked from others' practice. This is evidence-based medicine.

Nature/ontology (*phusis*) (see Chapter 1)

***Phusis*$_1$**: In very early contexts (pre-500 BCE) to say that biomedical events were produced by chance (a sort of magical serendipity devoid of divinity)—later (from Aristotle and the Hippocratic Writers onward), this becomes the position of a material/mechanistic view of nature's operation. These presuppositions are linked only metaphorically to divinity or metaphorically to a material cause—if D-□, then within this context, if D-◊ or E-◊, then to *phusis*$_3$.

***Phusis*$_2$**: That natural phenomena (including biomedical phenomena) were directly caused by a divinity, *phusis*$_2$ (sometimes also referred to as magic). The inexplicable are accepted by faith. They weigh in according to modality as D-□. Later philosophers may substitute reason as a substitute for "divinity."

***Phusis*$_3$**: The natural phenomena (including biomedical phenomena) are caused by unexplainable factors (also sometimes called magic). These proponents do not accept Dogmatic/logical accounts, as such. This is because of a rather high standard of epistemology (the careful scrutiny maxim). Because of this, they emphasize the modality of scientific study on the contingent side (◊) to various degrees.

Skepticism (see Chapter 4)

Skepticism$_1$: Falls in line with radical doubt. It contains the following elements—epistemologically, it embraces the selected faith maxim and explanatory-type β; the ontology is *phusis*$_2$; it is type oriented; and the modality is D-□/philosophical.

Skepticism$_2$: This form of skepticism describes the realm of the Dogmatists/logical practitioners. Epistemologically, it embraces the careful scrutiny maxim and explanatory-type α. The ontology is of *phusis*$_1$. Type and token are regarded equally. The modality is D-◊/philosophical.

Skepticism$_3$: The Empiricists reside here. Epistemologically, they embrace the extreme careful scrutiny maxim and explanatory-type β. Ontologically, they embrace *phusis*$_3$. They are token oriented, and the modality is E-◊/clinical.

Synthetic thinkers (See Chapter 5): Synthetic thinkers begin their explanation by acknowledging what has come before them. They look at the positive and accept those posits (often with a twist of their own). They also look at the negative (problems with earlier accounts) and use their own insights to solve these deficiencies. In this way the synthetic thinker tries to combine all the good that came before and combine it (with revisions) into his own vision that also satisfies deficiencies that exist in the previous accounts. In this project, all shortcomings of the past are solved by a comprehensive account. It is often the case that synthetic thinkers are also systematic thinkers. The complement to synthetic thinkers is the so-called *whole-cloth* thinkers who act as if they are addressing a particular project for the first time. They can be systemic or eristic, though generally they are eristic.

Systematic and eristic philosophers (see Chapter 5)

Eristic philosophers Eristic philosophers are interested in solving particular issues before them (the clinical approach). They are focused upon the token. Their ontological interest is in *phusis*$_3$. They tend toward *skepticism*$_3$ and adopt an extreme version of the epistemological careful scrutiny maxim. The Empiricists and Methodists fall into this category. They often assume in their discussion that they are addressing a problem for the first time (i.e., they do not contextualize their discussion via named predecessors as synthetic philosophers do). Many eristic philosophers fall into the *whole-cloth* approach because they eschew *historia* except as a clinical guidebook of what has worked in the past.

Systematic philosophers Synthetic philosophers are interested in a broad, comprehensive vision that seeks to explain an entire discipline. Because this is a theoretical exercise, these thinkers are inclined to types, the selected faith maxim, and aspire to D-□ as much as possible and to D-◊ when the data are not strong enough to create a necessary modality. In most cases, these philosophers are also synthetic thinkers, but in some cases, they claim to originate their grand vision from whole cloth.

Type/token (See Chapter 2)

Type: Refers to a general categorical description. In modal logic necessity, (□) only resides here because when dealing abstractly with groups, one can model the argument so that it follows without exception. Some biomedical practitioners believe that the practice of science is all about discovering and following general laws of nature. The ontology that follows is type driven. The philosophical practice embraces the ontology of *phusis*$_2$ via the epistemology of the selected faith maxim. Balanced type-token accounts embrace the ontology of *phusis*$_1$ and the epistemology of the careful scrutiny maxim—in which case the modality changes to moderately contingent ◊.

Token: Refers to individuals. In modal logic, discussion of individuals is highly contingent (◊), since individuals vary greatly. Some biomedical practitioners believe that the practice of science is all about confronting and recognizing the individuality of the *explanandum*—thus, the practice is firmly E (Empiricist-based). In an extreme case, advocates of the token as the sole object of investigation believe there are no truly common general classes, but all are *sui generis*. The ontology is *phusis*$_3$, and the epistemological stance is the careful scrutiny maxim.

Bibliography

Adkins, A. W. H. (1960) *Merit and Responsibility*. Oxford: Oxford University Press.

Aeschyli *Septem quae supersunt tragoedias*. (1972) Denys Page, ed. Oxford: Clarendon Press.

Aeschines *Epistulae*. (1896) F. Blass, ed. Leipzig: Teubner.

Aird, W. C. (2011) "Discovery of the Cardiovascular System: From Galen to William Harvey" *Journal of Thrombosis and Haemostasis* 9 (Supplement 1): 118–129.

Anonymous Parisian of Fuchs's *Anecdota medica* (1894) R. Fuchs, "Anecdota Medica Graeca," *Rheinisches Museum* 49: fragment 1.

Anonymus Londinensis (1893) *Anonymi Londinensis ex Aristotelis Iatricis Menoniis et aliis medicis eclogue*. H. Diels, ed. Berlin: G. Reimer Königlich Preussische Akademie der Wissenschaften.

_____. (1947) *The Medical Writings of Anonymus Londinensis*. W. H. S. Jones, ed. Cambridge: Cambridge University Press.

Aristotle. (1831) *Aristotelis Opera*. A. I. Bekker, ed. Berlin: G. Reimerum.

_____. *Historia Animalium*. (1965) tr. A. L. Peck. Cambridge, MA: Harvard University Press (Loeb).

_____. *Parts of Animals*. (1882) tr. W. Ogle. London: Kegan Paul, Trench & Co.

_____. *Parts of Animals, Movement of Animals, Progression of Animals*. (1968) Tr. A. L. Peck and E. S. Forster. Cambridge, MA: Harvard University Press (Loeb).

_____. *Aristotle's* De Partibus Animalium I and De Generatione Animalium I (with passages from II.3) (1972) tr. D. Balme. Oxford: Clarendon Press.

Arrian (1976) *Anabasis Alexandri*. Ed. and tr. P. A. Brunt. Cambridge: Cambridge University Press.

Ashmore, M. (1993) "The Theater of the Blind: Starring a Promethean Pranskter, a Prism, a Pocket, and a Piece of Wood" *Social Studies of Science* 23: 67–106.

Asmis, E. (1984) *Epicurus' Scientific Method*. Ithaca, NY: Cornell University Press.

Asper, M. (2013). "Explanation between Nature and Text: Ancient Greek Commentators on Science." *Studies in History & Philosophy of Science Part A* 44.1: 43–50.

Augustine (1942) *Against the Academicians*, ed. and tr. Mary Garvey. Milwaukee, WI: Marquette University Press.

Ayala, F. (1989) "Teleological Explanations" in Philosophy of Biology, M. Ruse, ed. (pp. 187–197). New York: Macmillan.

Azizi, M-H., T. Nayermouri, and F. Azizi (2008) "A Brief History of the Discovery of the Circulation of Blood in the Human Body" *Archive of Iranian Medicine* 11(3): 345–350.

Babb, J. and P. Littledog (1999) *Border Healing Woman: The Story of Jewel Babb as Told to Pat Littledog*. 2nd ed. Austin, TX: University of Texas Press.

Baker, M. A. (1972) "Influence of the Carotid Rete on Brain Temperatures in Cats Exposed to Hot Environments" *Journal of Physiology*. 220(3): 771–728.

Balaguer, M. (1999) "Review of *Mathematics as a Science of Patterns*" *Philosophia Mathematica* 7(3): 108–126.

Balme, D. (1961) "Aristotle's Use of Differentiae in Zoology" in *Aristote et les problems de method* (195–212), ed. S. Mansion. Louvain: Publications Universitaires.

_____. (1980) "Aristotle's Biology Was Not Essentialist" *Archiv für Geschichte Philosophie*:1–12.

_____. Tr. (1972) *Aristotle's* De Partibus Animalium I and De Generatione Animalium I (with passages from II.3). Oxford: Clarendon Press.

_____. (1987) "Teleology and Necessity" in *Philosophical Issues in Aristotle's Biology*, A. Gotthelf and J. Lennox, eds. (pp. 275–286) Cambridge: Cambridge University Press.

Baltzly, D. (2004) "Melissus and the Pluralists: A Philosophical Morality Play" *Diotima: Review of Philosophical Research*. 32: 32–46.

Bardong, K. (1942) Beiträge apply Hippokrates-und Galenforschung. Göttingen: Vandenhoeck & Ruprecht.

Barnes, J. (1982) *Aristotle*. Oxford: Oxford University Press.

Batens, D. (1989) "Natural Heuristics for Proof Construction: Part 1—Classical Propositional Logic" *Logique et Analyse* 32: 337–363.

Beardslee, D. (1918) *The Uses of ΦΥΣΙΣ in Fifth-Century Greek Literature*. Chicago: University of Chicago Press.

Belkin, G. (2014) *Death Before Dying: History, Medicine, and Brain Death*. Oxford: Oxford University Press.

Berkeley, G. (1710) *On the Principles of Human Knowledge*. Dublin: A. Rhames for J. Pepyat.

Berryman, S. (2002) "Galen and the Mechanical Philosophy" *Apeiron* 35(3): 235–254.

_____. (2009) *The Mechanical Hypothesis in Ancient Greek Natural Philosophy*. Cambridge: Cambridge University Press.

Birch, T. (1756) *The History of the Royal Society of London*. London: The Royal Society of London.

Bondio, M. G., ed. (2005) *Blood in History and Blood Histories*. Florence: Sismel-Edizioni del Galluzzo.

Booth, N. B. (1960) "Empedocles' Account of Breathing" *Journal of Hellenic Studies* 80: 10–15.

Bos, A. P. (2002) "Pneuma and Ether in Aristotle's Philosophy of Living Nature" *Modern Schoolman: A Quarterly Journal of Philosophy*. 79(4): 255–276.

_____. (2003) "Aristotle on the Etruscan Robbers: A Core Text of "Aristotelian Dualism." *Journal of the History of Philosophy*. 41(3): 289–306.

Boudon-Millot, V. (2005) "Art, Science and Conjecture, from Hippocrates to Plato and Aristotle" in P. J. Van Der Eijk, ed. *Hippocrates in Context: Papers Read at the XIth International Hippocrates Colloquium University of Newcastle upon Tyle 27-31 August 2002*: Leiden: Brill, 87–99.

Bourgey, L. and J. Jouanna, eds. (1975) *La collection hippocratique et son rôle dans l'histoire de la médecine*. Leiden: E. J. Brill.

Boyd, R., P. Gasper, and J. D. Trout. (1991) *The Philosophy of Science*. Cambridge, MA: MIT Press.

Boylan, M. (1980) "Henry More and the Spirit of Nature" *Journal of the History of Philosophy* 18(4): 395–405.

_____. (1981) "Mechanism and Teleology in Aristotle's Biology" *Apeiron* 15(2) 96–102.

_____. (1982) "The Digestive and 'Circulatory' Systems in Aristotle's Biology" *Journal of the History of Biology* 15(1): 89–118.

_____. (1983) *Method and Practice in Aristotle's Biology*. Lanham, MD and London: Rowman and Littlefield/UPA.

_____. (1984) "The Hippocratic and Galenic Challenges to Aristotle's Conception Theory" *Journal of the History of Biology* 15(1): 83–112.

_____. (1985) "The Place of Nature in Aristotle's Biology" *Apeiron* 19(1): 126–139.
_____. (1986a) "Galen's Conception Theory" *Journal of the History of Biology* 19(1): 44–77.
_____. (1986b) "Monadic and Systemic Teleology" in *Modern Problems in Teleology*, N. Rescher, ed. (pp. 15–25) Washington, DC: University Press of America.
_____. (2004a) "Aristotle's Biology" *The Internet Encyclopedia of Philosophy*: http://www.iep.utm.edu/aris-bio/[accessed February 27, 2015].
_____. (2004b) "Galen" *The Internet Encyclopedia of Philosophy*: http://www.iep.utm.edu/galen/[accessed February 27, 2015].
_____. (2004c) "Hippocrates" *The Internet Encyclopedia of Philosophy*: http://www.iep.utm.edu/hippocra/[accessed February 27, 2015].
_____. (2007) "Galen on the Blood, Pulse, and Arteries" *Journal of the History of Biology* 40(2): 207–230.
_____. (2008) *The Good, The True, and The Beautiful*. London: Bloomsbury.
_____. (2009) *Basic Ethics*. 2nd ed. Upper Saddle River, NJ: Prentice Hall.
_____. (2010) "Aristotle the Outsider" in *Philosophy: An Innovative Introduction*, M. Boylan and C. Johnson, eds. (pp. 62–71) Boulder, CO: Westview.
_____. (2012) "Ethical Limitations on Scientific Research" in *Scientific Freedom*, S. Giordano, J. Coggon, and M. Cappato, eds. (pp. 149–161) London: Bloomsbury.
_____. (2014a) "Health as Self-fulfillment" in *Medical Ethics* 2nd ed. M. Boylan, ed. (pp. 44–64) Oxford: Wiley-Blackwell.
_____. (2014b) "The Oath" in *Medical Ethics* 2nd ed. M. Boylan, ed. (pp. 65–66) Oxford: Wiley-Blackwell.
Boylan, M. and K. Brown (2002) *Genetic Engineering: Science and Ethics on the New Frontier*. Upper Saddle River, NJ: Prentice Hall.
Boyle, R. (1980 [1660–1691]) *Selected Philosophical Papers of Robert Boyle*. M. A. Stewart, ed. Manchester: University of Manchester Press.
Brain, P. (1986) *Galen on Bloodletting*. Cambridge: Cambridge University Press.
Bricot, T. (1494) *Corpus optimarum questionum super philosphiam Aristotelis*. Freiburg: Kilanus Piscator.
Brock, A. J. editor and translator (1916) *On the Natural Faculties*, London: Heineman (Loeb).
Brock, S. and E. Mares (2007) *Realism and Anti-Realsim*. Durham: Acumen Publishers.
Buchowski, M. (1988) "The Rationality of Magic" *Philosophy of the Social Sciences* 18: 509–518.
Carastro, M. (2006) *La cité des mages*. Grenoble: Jérôme Millon.
Carnap, R. (1950) *Logical Foundations of Probability*. Chicago: University of Chicago Press.
Celsus (1935) *De Medicina*. Tr. W.G. Spencer. 3 vols. Cambridge, MA: Harvard University Press (Loeb).
Celsus, A. C. (1995) *De medicina*. Ed. and tr. G. Serbat. Paris: Les Belles Lettres.
Chang, H-H. (2008) "Rationalizing Medicine and the Social Ambitions of Physicians in Classical Greece" *Journal of the History of Medicine and Allied Sciences*. 63(2): 217–244.
Chaucer, G. "The Squires Tale" in *The Works of Geffrey Chaucer*, 2nd ed. F.N. Robinson (1957) Boston: Houghton Mifflin Company, 89–92.
Churchland, P. (1981) "Eliminative Materialism and the Propositional Attitudes" *Journal of Philosophy* 78(2): 67–90.
Cicero (1923) ed. W.A. Falconer. *On Old Age, On Friendship, On Divination*. Cambridge, MA: Harvard University Press.
_____. (1927) ed. J. E. King. *Tusculan Disputations*. Cambridge, MA: Harvard University Press.
Clifford, W. K. (1901) "The Ethics of Belief" in *Lectures and Essays*. Vol. 2 London: Macmillan & Co.: 163–205.

Collins, D. (2003) “Nature, Cause, and Agency in Greek Magic” *Transactions of the American Philological Association* 133: 17–49.
Cooper, J. (1987) “Hypothetical Necessity and Natural Teleology” in *Philosophical Issues in Aristotle’s Biology*. A. Gotthelf and J. Lennox, eds. (pp. 243–274) Cambridge: Cambridge University Press.
_____. (1988) “Metaphysics in Aristotle’s Embryology” *Cambridge Philological Society Proceedings*. 214 (new series 34): 14–41.
Cooper, J. ed. (1997) *Plato: Complete Works*. Indianapolis, IN: Hackett.
Cosans, C. E. (1997) “Galen’s Critique of Rationalist and Empiricist Anatomy” *Journal of the History of Biology* 30: 35–54.
_____. (1998) “The Experimental Foundations of Galen’s Teleology” *Studies in the History and Philosophy of Science* 29A(1): 63–80.
Dancy, J. (2004) *Ethics Without Principles*. Oxford: Clarendon Press.
Darenberg, C. V. (1869) *Ètat de la medicine entre Homère et Hipocrate*. Paris: Didler.
Davidoff, F., B. Hayne, D. Sackett, and R. Smith (1995) “Evidence-Based Medicine” *British Medical Journal* 310: 1085–1086.
Dean-Jones, L. (1992) “The Politics of Pleasure: Female Sexual Appetite in the Hippocratic Corpus” in *Discourses of Sexuality from Aristotle to AIDS*. D. C. Stanton, ed. (pp. 48–77) Ann Arbor: University of Michigan Press.
Deichgräber, K. (1965) *Die griechische Empirikerschule*, 2nd ed. Berlin: Weidmann.
DeLacy, P. tr. and ed. (1981) *On the Doctrines of Hippocrates and Plato. Corpus Medicorum Graecorum*. Vol. 4, pt. 1, fasc. 2. Berlin: Academie-Verlag.Demont, P. (1978) “Remarques sur le sense de τρέφω” *Revue des Études Grecques* 91: 358–384.
_____. (1998) “Note sûr les premiers emplois ἠθικός chez Aristote. Le Sentiment d’Amité et les transactions de gré à gré,” *Actes du Colloque International ‘Entre Public et Privé en Grèce Ancienne. Ktema* 23: 81–92.
_____. (2005) “About Philosophy and Humoural Medicine” in *Hippocrates in Context: Papers Read at the XI International Hippocrates Colloquium University of Newcastle Upon Tyne 27–31 August 2002*, P. J. van der Eijk, ed. (pp. 271–286) Leiden and Boston: Brill.
_____. (2013) “The Causes of the Athenian Plague and Thucydides” in A. Tsakmakis and M. Tamiolaki, eds. (pp. 73–90) *Thucydides Between History and Literature*. Boston and Berlin: Walter de Gruyter.
Demont, P. and J. Jouanna (1981) “Le Sens D’ἰχώρ Chez Homere (*Iliade* V. v. 340 et 416) et Eschyle (*Agamemnon*, v. 1480) en Relation Avec Les Emplois Du Mot Dans La *Collection Hippocratique*” *Revue des Études Anciennes* 83: 197–209.
Descartes, R (1897) *Oeuvres de Descartes*, eds. Adams, C. and Paul Tannery, 11 vols. Paris: J. Vrin.
_____. (1911 [1649]). *The Passions of the Soul*, tr. E. S. Haldane and G. R. T. Ross. Cambridge: Cambridge University Press.
_____. (1911 [1637]) *Discourse on Method*, tr. and edited by E. S. Haldane and G. R. T. Ross. Cambridge: Cambridge University Press.
Dicker, G. (2013) *Descartes: An Analytical and Historical Introduction*. 2nd ed. New York: Oxford University Press.
Dicks, D. R. (1970) *Early Greek Astronomy to Aristotle*. 2 vols. Ithaca, NY: Cornell University Press.
Diels, H., ed. (1893) *Anonymi Londinensis ex Aristotelis Iatricis Menoniis et aliis medicis eclogue*. Berlin: G. Reimer Königlich Preussische Akademie der Wissenschaften.
Diels, H. (1952) *Die Fragmente der Vorsokratiker*. 6th edition with revisions by W. Kranz. Berlin: Weidman.

Diocles of Carystus. (2000 [4th century BC]) *Diocles of Carystus: A Collection of the Fragments with Translation and Commentary*, volume 1, edited and translated by P. van der Eijk (Leiden: Brill).
Diocles of Carystus. (2001 [4th century BC]) *Diocles of Carystus: A Collection of the Fragments With Translation and Commentary*, volume 2, edited and translated by P. van der Eijk (Leiden: Brill).
Diogenes Laertius. (1925) *Lives of Eminent Philosophers*. tr. R. D. Hicks. London: Heinemann, Ltd.
Dobson, M. (2013) *The Story of Medicine: From Bloodletting to Biotechnology*. London: Quercus.
Dodds, E. R. (1951). *The Greeks and the Irrational*. Berkeley, CA: University of California Press.
Doney, W., ed. (1987) *Eternal Truths and the Cartesian Circle: A Collection of Studies* New York: Garland.
Dreyer, J. L. E. (1906) *History of the Planetary Systems from Thales to Kepler*. Cambridge: Cambridge University Press.
Ducatillon, J. (1977). *Polémiques dans la Collection Hippocratique*. Paris: H. Champion.
Duminil, M-P. (1983) *Le sang, les vaisseaux, le Coeur dans la collection hippocratique*. Paris, Les Belles Lettres.
Edelstein, E.J. and L. Edelstein (1945 [rpt. 1998]) *Asclepius: Collection and Interpretation of the Testimones*, 2 vols. Baltimore: Johns Hopkins University Press.
Edelstein, Ludwig (1929) "On the Sacred Disease" *Archiv fúr Geschichte der Philosophie* 22: 302.
_____. (1931) Περὶ ἀέρον und die Sammlung der Hippokratischen Schriften" *Problemata* 4. Berlin: Weidmann.
_____. (1939) "The Genuine Works of Hippocrates" *Bulletin of the History of Medicine* 7(1): 236–248.
_____. (1967) *Ancient Medicine*. Baltimore, MD: Johns Hopkins University Press.
Ellinger, Tage U. H. (1952) *Hippocrates on Intercourse and Pregnancy*. New York: Henry Schuman.
Epicurus, (2003) "On Nature" in *Epicurus, Opera, Fragmmenti, Testimonianze sulla sua vita/Epicuro*. Intro Gabriele Giannatoni, Ettore Bignone, tr. Rome: Editori Laterza.
Eristratus (1988) Ed. I. Garofalo. *Erasistrati Fragmenta*. Pisa: Giardini.
Euripides, *Fabvlae*, ed. G. Murray (1909), volume 3. Oxford: Clarendon Press.
Ferejohn, M. (1991) *The Origins of Aristotelian Science*. New Haven, CT: Yale University Press.
Fisher, B. and C. Redmond (May 19, 1994) "Fraud in Breast-Cancer Trials" *New England Journal of Medicine* 330: 1458–1462.
Field, H. (2009) "Epistemology without Metaphysics" *Philosophical Studies* 143(2): 249–290.
Fitts, D. (1947). *Greek Plays in Modern Translation*. New York: Dial Press.
Flemming, R. (2007) "Galen's Imperial Order of Knowledge" in *Ordering Knowledge in the Roman Empire*, J. König and T.Whitmarsh, eds. (pp. 241–277) Cambridge: Cambridge University Press.
_____. (2008) "Commentary" in Hankinson, R. J., ed. (pp. 323–354) *The Cambridge Companion to Galen*. Cambridge: Cambridge University Press.
Fodor, J. A. (1981) "The Mind-Body Problem" *Scientific American* 244: 114–123.
Forrester, J.M. (1954) "An Experiment of Galen's Repeated" *Proceedings of the Royal Society of Medicine* 47(4) 241–244.
Francis, S. (2011) " 'Under the Influence': The Physiology and Therapeutics of Akrasia in Aristotle's *Ethics*" *Classical Quarterly* 61(1): 143–171.

Fraser, P. M. (1969) "The Career of Erasistratus of Ceos," *Rendiconti*, Classe di Lettere Instituto Lombardy—Academia de Scienze e Lettere, 103. Milan: 518–537.
Frede, M. (1980) "The Original Notion of Cause" in J. Barnes, ed. *et al., Doubt and Dogmaticism*. Oxford: Oxford University Press.
_____. (1981) "On Galen's Epistemology" in *Galen's Problems and Prospects*, V. Nutton, ed. (pp. 65–86) London: Welcome Institute.
_____. (1985) *Galen: Three Treatises on the Nature of Science*. Tr. M. Frede and R. Walzer. Indianapolis: Hackett.
_____. (1986) "Philosophy and Medicine in Antiquity" in *Human Nature and Natural Knowledge: Essays Presented to Marjorie Grene on the Occasion of Her Seventy-Fifth Birthday*, A. Donagan and A. Perovich, eds. (pp. 211–232) Dordrecht: D. Reidel.
_____. (1987) "The Ancient Empiricists" in *Essays in Ancient Philosophy*. (pp. 243–260) Oxford: Clarendon Press.
Fredrich, C. J. (1899) *Hippokratische Untersuchugen*, Heft 15. Berlin: Weidmannsche Buchhandlung.
Fuchs, R. (1894) "Anecdota Medica Graeca" *Rheinisches Museum* 49.
Furley, D. (1957) "Empedocles and the Clepsydra: A Theory of Breathing" *Journal of Hellenic Studies* 77: 31–34.
_____ and J.S. Wilke (1984) *Galen: On Respiration and the Arteries*. Princeton: Princeton University Press.
Furth, M. (1988). *Substance, Form and Psyche: An Aristotelian Metaphysics*. Cambridge: Cambridge University Press.
Galen (1821–1833) *Opera Omnia*, edited and translated by C. G. Kühn. Leipzig: Prostat in officina libraria Car. Cnoblochii.
_____. (1964 [1821–1833]) *Opera Omnia*, edited by C. G. Kühn. 20 vols. Leipzig, rpt, Hildesheim: Olms.
_____. (1854) *De usu partium*, edited and translated by C. Daremberg. Paris: Ballière, 1854.
_____. (1874) *De Placitis Hippocratis et Platonis*, edited by I. Mueller. Lipsiae: Teubner.
_____. (1907–1909) *De usu partium* VI, critical edition edited by G. Helmreich. Leipzig: Teubner
_____. (1916) *On the Natural Faculties*, edited and translated by A. J. Brock. London: Heineman (Loeb).
_____. (1944) *On Medical Experience*. Oxford: Clarendon Press.
_____. (1956) *On Anatomical Procedures*, translated by C. Singer. Oxford: Oxford University Press.
_____. (1968) *Galen: On the Usefulness of the Parts of the Body*, translated by M. T. May. Ithaca, NY: Cornell University Press.
_____. (1981) *On the Doctrines of Hippocrates and Plato*, translated and edited by P. DeLacy. *Corpus Medicorum Graecorum*. Vol. 4, pt. 1, fasc. 2. Berlin: Academie-Verlag.
_____. (1984) *Galen: On Respiration and the Arteries*, translated by D. Furley and J. S. Wilke. Princeton: Princeton University Press.
_____. (1985) *Three Treatises on the Nature of Science*, translated by M. Frede and R. Walzer. Indianapolis, IN: Hackett.
_____. (1986) *Galen on Bloodletting*, edited and translated by P. Brain. Cambridge: Cambridge University Press.
_____. (1991) *Galen on the Therapeutic Method, Books I and II*, translated by. R.J. Hankinson. Oxford: Clarendon Press.
_____. (1998) *Galen: On Antecedent Causes*, translated by R. J. Hankinson. Cambridge: Cambridge University Press.
_____. (1997) *Galen: Selected Works*, edited and translated by P.N. Singer. Oxford: Oxford University Press.

_____. (2011) *Method of Medicine*, edited and translated by I. Johnston and G. H. R. Horsley. Cambridge, MA: Harvard (Loeb).

Gardner, M. (1957) *Fads and Fallacies in the Name of Science*. NY: Dover.

Garofalo, Ivan, ed. (1988) *Erasistrati Fragmenta*. Pisa: Giardini.

Gewirth, A. (1941) "The Cartesian Circle" *Philosophical Review* 50: 368–395.

_____. (1970) "The Cartesian Circle Reconsidered" *The Journal of Philosophy* 67: 668–685.

Gill, C. (1977). "Galen vs. Chrysippus on the Tripartite Psyche in '*Timaeus*' in *Interpreting the 'Timeaeus-Critias,'* T. Calvo, (pp. 69–72) Academia: Sankt Augustin.

_____. (1983) "Did Chrysippus Understand Medea?" *Phronesis* 28: 136–149.

_____. (2009) "Helenistic and Roman Philosophy" *Phronesis* 54(3): 286–296.

Gimbel, S. (2011) *Exploring the Scientific Method: Cases and Questions*. Chicago: University of Chicago Press.

Giorgianni, F. ed. (2006) *Hippokrates, Über die Natur des Kindes* (De genitura und De natura pueri). Wiesbaden: Dr. Ludwig Reichert Verlag.

Gomperz, T. (1922) *Griechische Denker: Eine Geschichte der antiken Philosophie*. Berlin: W. de Gruyter.

Gooding, D. C. (1990) *Experiment and the Making of Meaning*. Dordrecht: Kluwer.

Goodman, N. (1955). *Fact, Fiction, and Forecast*. Cambridge, MA: Harvard University Press.

Gordon, R. (1987) "Aelian's Peony: The Location of Magic in Graeco-Roman Tradition" *Comparative Criticism* 9: 59–95.

_____. (1999a) "Imagining Greek and Roman Magic" in *Witchcraft and Magic in Europe*, volume 2: Ancient Greece and Rome. V. Flint, R. Gordon, G. Luck, and D. Ogden, eds. (pp. 159–265) London: The Athlone Press.

_____. (1999b) "What's in a List? Listing in Greek and Graeco-Roman Malign Magical Texts" in *The World of Ancient Magic*. D.R. Jordan, H. Montgomery, and E. Thomassen, eds. (pp. 239–278) Norway: Norwegian Institute at Athens.

_____. (2002) "Shaping the Text: Innovation and Authority in Graeco-Egyptian Malign Magic" in *Κυκεον: Studies in Honour of H. S. Versnel*. H. F. J. Horstmanshoff, H.W. Singor, F. T. Van Straten, and J. H. M. Strubbe, eds. (pp. 69–112) Leiden: Brill.

Gotthelf, A., ed. (1985) *Aristotle on Nature and Living Things: Philosophical and Historical Essays in Honor of David M. Balme's Seventieth Birthday*. Pittsburgh, PA: Mathesis Press.

_____. (2012) *Teleology, First Principles, and Scientific Method in Aristotle's Biology*. Oxford: Oxford University Press.

Gourevitch, D. (1989) "'L'Anonyme' De Londres et la Médecine d'Italie Du Sud" *History and Philosophy of the Life Sciences* 11(2): 237–251.

Grams, L. (2009) "Medical Theory in Plato's *Timaeus*" *Rhizai: A Journal for Ancient Philosophy and Science* 6(2): 161–192.

Grapow, H., ed., et al. (1954–1973) *Grundriss der Medizin der alten Ägypter* 9 vol. Berlin: Akademie Verlag: IV, 64.

Graz, B., et al. (January 2010) "Argemone Mexicana Decoction versus Artesunate—Artesunate-Amodiaquine for the Management of Malaria I Mali: Policy and Public Health Implications" *Transactions of the Royal Society of Tropical Medicine and Hygiene*. 104(1): 33–41.

Grene, M. (1963) *A Portrait of Aristotle*. Chicago: University of Chicago Press.

Grice, P. (1969) "Utterer's Meaning and Intentions" *Philosophical Review* 78: 147–177.

Gundert, B. (1992) "Parts and Their Roles in Hippocratic Medicine" *Isis* 83(3): 453–465.

Gutherie, W. K. C. (1962) *A History of Greek Philosophy*. Vol. 1 Cambridge: Cambridge University Press.

Hacking, I. (2001) *An Introduction to Probability and Inductive Logic*. Cambridge: Cambridge University Press.

Hall, T. S. (1975). "Euripus; or, the Ebb and Flow of the Blood" *Journal of the History of Biology* 8(2): 321–350.
Hankinson, R. J. (1987) "Causes and Empiricism" *Phronesis* 32(3): 329–348.
_____. (1988) "Galen Explains the Elephant" in *Philosophy and Biology*, M. Matthen and B. Linsky, eds. (pp. 135–158) Calgary, Alberta: The University of Calgary Press.
_____. (1990) "Saving the Phenomena" *Phronesis* 35: 194–215.
_____. (1991a) "Galen's Anatomy of the Soul" *Phronesis* 36(3): 197–233.
_____. (1991b) *Galen on the Therapeutic Method, Books I and II*. Tr. R.J. Hankinson. Oxford: Clarendon Press.
_____. (1992) "Galen's Philosophical Eclecticism" *Aufstieg und Niedergang der rõmischen Welt* II 36.5.
_____. (1993) "Actions and Passions" in *Passions and Perception*,. J. Brunschwig and M. Nussbaum, eds. (pp. 184–222) Cambridge: Cambridge University Press.
_____. (1994) "Galen's Concept of Scientific Progress" in *Aufstieg und Niedergang der römischen Welt*. 372(2), W. Hasse and H. Temporini, eds. (pp. 1775–1789) Berlin: Walter De Gruyter.
_____. (1995a) "Pollution and Infection: An Hypothesis Stillborn" *Apeiron* 28: 25–65.
_____. (1995b) *The Sceptics: The Arguments of the Philosophers*. London: Routledge.
_____. (1995c) "The Growth of Medical Empiricism" in *Knowledge and the Scholarly Medical Traditions*, D. Bates, ed. (pp. 41–59) Cambridge: Cambridge University Press.
_____. (1998a) "Magic, Religion, and Science: Divine and Human in the Hippocratic Corpus" *Apeiron* 31(1): 1–34.
_____. (1998b) *Cause and Explanation in Ancient Greek Thought*. Oxford: Clarendon Press.
_____. (1998c) *Galen: On Antecedent Causes*, edited and translated by R. J. Hankinson. Cambridge: Cambridge University Press.
_____. (2002) *Cause and Explanation in Ancient Greek Thought*. Oxford: Clarendon.
_____. (2003) "Stoicism and Medicine" in *The Cambridge Companion to the Stoics*, B. Inwood, ed. (pp. 295–309) Cambridge: Cambridge University Press.
_____. (2005) "Explanation and Causation" in *The Cambridge History of Hellenistic Philosophy*. A. Keimpe, ed. (pp. 479–512) Cambridge: Cambridge University Press.
_____. (2008) *The Cambridge Companion to Galen*. Cambridge: Cambridge University Press.
Harris, C. and R. Schiller (1973) *The Heart and the Vascular System in Ancient Greek Medicine*. Oxford: Clarendon Press.
Harvey, W. (1628) *Exercitatio Anatomica de Motu Cordis et Anguinis in Animalibus*. Frankfurt: W. Fitzer/ rpt London Sydenham Society.
_____. (1957) *De motu cordis: Movement of the Heart and Blood in Animals*, translated by K. J. Franklin. Oxford: Published for the Royal College of Physicians by Basil Blackwell.
Heinimann, F. (1945) *Nomos und Physis: Herkunft und Bedeutung einer Antithese im griechischen Denken des 5. Jarhunderts*. Basel: Reinhardt.
Held, B. (1999) "The Question for Postmodern Therapists: To Be or Not To Be Theoretical" *Canadian Journal of Continental Philosophy* 3(1): 5–25.
Hempel, C. (1945) "Studies in the Logic of Confirmation, Parts I and II" *Mind* 54: 1–26, 97–121.
Henry, D. (2006) "Understanding Aristotle's Reproductive Hylomorphism." *Apeiron* 39(3): 257–287.

_____. (2008) "Aristotle on Definition" *Journal of the History of Philosophy* 46(3): 478–480.

Herodoti, *Historiae* (1908) C. Hude, ed. 2 vol. Oxford: Oxford University Press.

Herophilus (1989) *Herophilus: The Art of Medicine in Early Alexandria*. H. von Staden, ed. Cambridge: Cambridge University Press.

Hesiod (1990) *Theogonia, Opera et Dies, Scutum, Fragmenta Selecta*. F. Solmsen, R. Merkelbach, and M. L. West, eds. Oxford: Clarendon Press.

Hippocrate, Oeuvres Complètes d'. (1839–61), translated and edited by E. Littré. 10 vol. Paris: J. B. Billière.

_____. (1923–1931) *Hippocrates, selected treatises*, translated by W. H. S. Jones. 4 vols. Cambridge, MA: Harvard University Press, supplemented by W. Smith (vol. 7, 1994) and by P. Potter (vols. 5, 6, 8, 9, and 10 1988–2012) (Loeb).

_____. (1970) *Hippocrate, De la generation, De la nature de l'enfant, et al.* vol. XI. R. Joly, ed. Paris: Budé.

_____. (1981) *On Generation, On the Nature of the Child, and Diseases IV*, I. Lonie, ed. Berlin and New York: DeGruyter.

_____. (2006) *Hippokrates, Über die Natur des Kindes* (De genitura und De natura pueri), F. Giorgianni, ed. Wiesbaden: Dr. Ludwig Reichert Verlag.

Hoffman, R., E. J. Benz, et al. (2012) *Hematology: Basic Principles and Practice*. 6th edition. London: Churchill Livingstone.

Homer, *Iliad*. (1951) Tr. R. Lattimore. Chicago: University of Chicago Press.

_____. *Odyssey*. (1965) Tr. R. Lattimore. NY: Harper and Row.

Hong, H. (2014) *The Theory of Chinese Medicine*. London: Imperial College Press.

Horowitz, M. C. (1976) "Aristotle and Women," *Journal of the History of Biology* 9: 186–213.

Hudnall, S. D. (2012) *Hematology: A Pathophysiologic Approach*. Philadelphia: Mosby-Elsevier.

Hudson, R. G. (1997) "Classical Physics and Early Quantum Theory: A Legitimate Case of Theoretical Underdeterminism" *Synthese* 110(2): 217–256.

Hull, D. (1974) *Philosophy of Biological Science*. Englewood Cliffs, NJ: Prentice Hall.

Hume, D. (1977 [1748]) *An Enquiry Concerning Human Understanding*, Eric Steinberg, ed. Indianapolis, IN: Hackett.

Hüttemann, A. (2000) "Descartes' Kritik an den realen Qualitäten: das Beispiel der Schwere" *Archiv für Geschichte der Philosophie* 83: 24–44.

Huxley, T. H. (1880) "On Certain Errors Respecting the Structure of the Heart Attributed to Aristotle" *Nature* 21: 1–5.

Hwa, C., A. Sebastian, and W. C. Aird, (2005) "Endothelial Biomedicine: Its Status as an Interdisciplinary Field, Its Progress as a Basic Science, and Its Translational Bench-to-Bedside Gap" *Endothelial Biomedicine* 12(1): 1–13.

Ilberg, J. (1897) "Über die Schriftstellerei des Klaudios Galenos," *Rheinisches Museum für Philologie* ns 52:165–196.

James, W. (1897) "The Sentiment of Rationality" in *The Will to Believe and Other Essays* (pp. 63–110). New York: Longmans, Green, & Co.

Jenkins, M. (2011) "Socrates and the Theaetetus" *Logos and Episteme: An International Journal of Epistemology* 2(1): 51–68.

Joly, R. ed. (1970) *Hippocrate, De la generation, De la nature de l'enfant, et al.* vol. XI. Paris: Budé.

Jones, W.H.S., ed. (1947) *The Medical Writings of Anonymus Londinensis*. Cambridge: Cambridge University Press.

Jori, A. (2005) "Blut und Leben bei Aristoteles" in *Blood in History and Blood Histories*, M. G. Bondio, ed. (pp. 19–38) Florence: Sismel-Edizioni del Galluzzo.

Jouanna, J. (1974) *Hippocrate: Pour une Archéollgie de L'École de Cnide*. Paris: Belles Lettres.

_____. (1980) "Politique et Medicine. La Problématique du Changement dans RMA et Chez Thucydide" in *Hippocratica Actes du Colloque International Hippocratique de Paris*, M. D. Grmek, ed. (pp. 299–319) Paris: Éditions du Centre National de la Recherche Scientifique.

_____. (2012a) "Egyptian Medicine and Greek Medicine" in *Greek Medicine from Hippocrates to Galen*, P. van der Eijk, ed. and N. Allies, tr. (pp. 3–20) Leiden: Brill.

_____. (2012b) "Wine and Medicine in Ancient Greece" in *Greek Medicine from Hippocrates to Galeni* P. van der Eijk, ed. and N. Allies, tr. (pp. 173–193) Leiden: Brill.

Jouanna, J. and P. Demont (1981) "Le Sens D'ἰχώρ Chez Homere (*Iliade* V. v. 340 et 416) et Eschyle (*Agamemnon*, v. 1480) en Relation Avec Les Emplois Du Mot Dans La *Collection Hippocratique*" *Revue des Études Anciennes* 83: 197–209.

Kahn, C. (1979) *The Art and Thought of Heraclitus: An Edition of the Fragments with Translation and Commentary*. Cambridge: Cambridge University Press.

Kant, I. (1781, 1787, rpt 1968) *Kritik der reinen Vernunft*. Berlin: Walter de Gruyter.

Karshren, E. (2011) "Thought, Utterance, Power: Toward a Rhetoric of Magic" *Philosophy and Rhetoric* 44(1): 52–71.

Kirk, G. S., J. E. Raven, and M. Schofield (1983) *The Pre Socratic Philosophers*. Cambridge: Cambridge University Press.

Klibansky, R. E., E. Panowsky, and F. Saxl (1964) *Saturn and Melancholy*. London: Nelson.

Kuhn, T. S. (1957) *The Copernican Revolution*. Cambridge, MA: Harvard University Press.

_____. (1962) *The Structure of Scientific Revolutions*, 2nd ed. Chicago: The University of Chicago Press.

Kullmann, W. (1974) *Wissenschaft und Methode: Interpretationen zur Aristotelischen Theorie der Naturwissenschaft*. Berlin: Walter de Gruyter.

Kuriyama, S. (1995) "Interpreting the History of Bloodletting" *Journal of the History of Medicine and Allied Sciences* 50: 11–46.

_____. (2002) *The Expressiveness of the Body and the Divergence of Greek and Chinese Medicine*. Cambridge, MA: Zone Books.

Laertius, D. (1964) *Diogenis Laertii Vitae Philosophorum*, H. S. Long, ed. Oxford: Clarendon Press.

Lakatos, I. (1970) "Falsification and the Methodology of Scientific Research Programmes" in *Criticism and the Growth of Knowledge*, I. Imre Lakatos and A. Musgrave, eds. (pp. 170–196) Cambridge: Cambridge University Press.

Lange, M. (1994) "Scientific Realism and Components: The Case of Classical Astronomy" *Monist* 77(1): 111–127.

Lattimore, R. tr. (1951) Homer, *Iliad*. Chicago: University of Chicago Press.

_____. *Odyssey*. (1965). New York: Harper and Row.

Lefkowitz, M. (2005) *Women's Life in Greece and Rome*. 3rd ed. Baltimore, MD: Johns Hopkins University Press.

Leith, D. (2008) "The *diatritus* and Therapy in Graeco-Roman Medicine" *Classical Quarterly* 58: 58–60.

Lesky, E. (1950) *Die Zeugungs und Vererbungslehre der Antike und ihre Nachwirkung*. Mainz: F. Steiner.

Levin, S. B. (2014) *Plato's Rivalry with Medicine: A Struggle and Its Dissolution*. Oxford: Oxford University Press.

Limet, R. (2010) "D'Hippocrate À Harvey: Vingt Siècles D'histoire de la Circulation" *Revue Médicale de Liège* 65(10): 562–568.

Littré, E. translator and editor (1839–1861) *Hippocrate, Oeuvres Complètes* 10 vol. Paris: J. B. Billière.

Lloyd, G. E. R. (1963) "Who Is Attacked in Ancient Medicine?" *Phronesis* 8(1): 108–126.

_____. (1966) *Polarity and Analogy*. Cambridge: Cambridge University Press.
_____. (1973) *Greek Science After Aristotle*. New York: Norton.
_____. (1975) "The Hippocratic Question" *The Classical Quarterly* 25(2): 171–192.
_____. (1979) *Magic, Reason, and Experience*. Cambridge: Cambridge University Press.
_____. (1987) *The Revolutions of Wisdom*. Berkeley, CA: University of California Press.
_____. (2007) "*Pneuma* between Body and Soul" *Journal of the Royal Anthropological Institute* (n.s. 13.1): S135–146.
_____. (2008) "Galen and His Contemporaries" in *The Cambridge Companion to Galen*, Hankinson, ed. (pp. 34–48) Cambridge: Cambridge University Press.
Lock, M. (2001) *Twice Dead: Organ Transplants and the Reinvention of Death*. Berkeley: University of California Press.
"London Papyrus" (British Museum 10059, 37, cf. 158).
Lones, T. E. (1912) *Aristotle's Researches in Natural Science*. London: West, Newman & Co.
Long, A. A. (1986) *Hellenistic Philosophy: Stoics, Epicureans, Sceptics*. 2nd ed. Berkeley, CA: University of California Press.
Long, A. A. and D. N. Sedley (1987) *The Hellenistic Philosophers*. Cambridge: Cambridge University Press.
Longino, H. (2002) *The Fate of Knowledge*. Princeton: Princeton University Press.
Longrigg, J. (1993) *Greek Rational Medicine: Philosophy and Medicine from Alcmaeon to the Alexandrians*. London: Routledge.
_____. (1995) "Medicine and the Lyceum" in *Ancient Medicine in Its Socio-Cultural Context: Papers Read at the Congress Held at Leiden University, 13–15 April, 1992*, P. van der Eijk, H. F. J. Horstmanshoff, and P. H. Schrijvers, eds., vol. 2. (pp. 431–446) Amsterdam: Rodopi.
_____. (1998) *Greek Medicine from the Heroic to the Hellenistic Age: A Source Book*. London: Routledge.
Lonie, I. ed. (1981) *On Generation, On the Nature of the Child, and Diseases IV*, Berlin: DeGruyter.
Lorenz, K. (1996) *On Aggression*. 2nd ed. London: Routledge.
_____. (2002) *King Solomon's Ring*. 2nd ed. London: Routledge.
Magnus, D. (1997) "Heuristics and Biases in Evolutionary Biology Heuristics" *Biology and Philosophy* 12(1): 21–38.
Manetti, D. (2013) "Levels of Authorial Presence in *Anonymus Londiniensis* (P. Brit. Libr. Inv. 137)" *Trends in Classics* 5(1): 159–178.
Mani, N. (1959) "Die historischen Grundlagen der Legerforschung," I Tiel *Die Vorstellungen Über Anatomie, Physiologie und Pathologie der Leber in der Antike* Basel/Stuttgart: Baseler Veröffentlichungen zur Geschichte der Medizin und der Biologie: 67–70.
Marganne, M.-H (1980) "Sur l'origine hippocratique des concepts de revulsion et de derivation" *L'Antiquité* 49: 115–130.
Martini, D. (1962) "Considerazione sulla dottrina dello pneuma in Galeno," in *Pagine di Storia della Medicina*, C. Castellani, ed. Rome: Artigrafiche E. Cossidente.
Matthen, M. (1988) "Empiricism and Ontology in Ancient Medicine" *Apeiron* 21(2): 99–121.
Mattern, S. P. (2013) *The Prince of Medicine: Galen in the Roman Empire*. Oxford: Oxford University Press.
Mayhew, R. (2004) *The Female in Aristotle's Biology*. Chicago: University of Chicago Press.
McClure, L. (1999) *Spoken Like a Woman*. Princeton, NJ: Princeton University Press, 1999.
Mejer, J. (2002) "Eudemus and the History of Science" in *Eudemus of Rhodes*, M. I. Bodnár and W. W. Fortenbaugh, eds. *Rutgers University Studies in Classical Humanities*, vol. XI (pp 252–253) New Brunswick: Transaction Publishers.

Michael of Ephesus (1903) *In Libros, de Generatione Animalium commentaria*, M. Hayduck, ed. Berlin: G. Reimeri.
Mill, J. S. (1843) *A System of Logic: Ratiocinative and Inductive*. London: John W. Parker.
More, H. (1671) *Enchiridion Metaphysicum; sive, de rebus incorporeia succincta & luculenta dissertation*. London: E. Flesher and G. Mordem.
Morrison, K. (1999) "Origins and Influences: Answering Held's Call to Realism" *Canadian Journal of Continental Philosophy* 3(1): 27–41.
Morsink, J. (1982) *Aristotle on the Generation of Animals*. Lanham, MD: University Press of America.
Mucz, M. (2011) *Baba's Kitchen Medicines*. Edmonton: University of Alberta Press.
Mueller, I. ed. (1874) *De Placitis Hippocratis et Platonis*. Lipsiae: Teubner.
Mueller, I. (1981) *Philosophy of Mathematics and Deductive Structure in Euclid's Elements*. Cambridge, MA: MIT Press.
National Library of Medicine (2002) "Summaries for Patients: Do Women with Heart Attacks Get Fewer Procedures than Men?" *Annals of Internal Medicine*. 137(6): 126.
Nawer, T. (2013) "Knowledge and True Belief at *Theaetetus* 201 a-c" *British Journal for the History of Philosophy* 21(6): 1052–1070.
Needham, J. (1959) *History of Embryology*. 2nd edition. New York: Abelard-Schuman.
Nickel, D. (2002) "Hippokratisches bei Praxagoras von Kos?" in *Hippocrates in Context: Papers Read at the XI International Hippocrates Colloquium, University of Newcastle on Thyme, 27–31 August 2002*, P. J. van der Eijk, ed. (pp. 315–323) Leiden: Brill.
Noll, R. (1919) "Zu Galens Schrift ει κατα φυσιν . . ." *Berliner Philologischer Wochenschrift* 33: 1246–1247.
Novotny, J. A., S. K. Gebauer, and D. J. Baer (2012) "Discrepancy Between the Atwater Factor Predicted and Empirically Measured Energy Values of Almonds in Human Diets" *American Journal of Clinical Nutrition*. 96(2): 296–301.
Nussbaum, M. (1978) *Aristotle's De motu animalium*. Princeton, NJ: Princeton University Press.
Nutton, Vivian (1969) "The Doctor and the Oracle" *Revue Belge de Philologie* 47: 37–48.
_____, ed. (1981) *Galen's Problems and Prospects*. London: Welcome Institute.
_____. (1985) "Murders and Miracle Cures: Lay Attitudes Towards Medicine in Classical Antiquity" in *Patients and Practitioners*, R. Porter, ed. (pp. 23–53) Cambridge: Cambridge University Press.
_____. (1993) "Roman Medicine: Tradition, Confrontation, Assimilation." in H. Temporini and W. Haase eds. (1974—) *Aufstieg und Niedergang der Römischen Welt: Geschichte und Kultur Roms im Spiegel neueren Forschung*. II. 37.1 (pp. 49–78) Berlin: De Gruyter.
_____. (2005) "The Fatal Embrace: Galen and the History of Ancient Medicine" *Science in Context*. 18(1L): 111–121.
_____. (2008) "Greek Medical Astrology and the Boundaries of Medicine" in *Astro-Medicine: Astrology and Medicine, East and West*, A. Akasoy, C. Burnett, and R. Yoeli-Tlalim, eds. (pp. 17–31) Florence: SISMEL—Edizioni del Galluzzo: 17–31.
_____. (2013) *Ancient Medicine*. 2nd edition. London and New York: Routledge.
_____. (forthcoming) "Galen Avoiding Distress" translation in P. N. Singer, ed. *Galen: Psychological Writings*. Cambridge: Cambridge University Press.
_____. (forthcoming) "Le Sang et l'Air" unpublished manuscript.
O'Brien, D. (1970) "The Effect of a Simile: Empedocles' Theories of Seeing and Breathing" *The Journal of Hellenic Studies* 90: 140–179.

Ogle, W. tr. (1882) *Parts of Animals*. London: Kegan Paul, Trench & Co.

Olson, S. D. (1995) *Blood and Iron*. Leiden: Brill.

Page, D. ed. (1972) Aeschyli *Septem quae supersunt tragoedias*. Oxford: Clarendon Press.

Paley, W. (1867) *Natural Theology: Or Evidences of the Existence and Attributes of the Deity Collected from the Appearances of Nature*. Boston: Gould and Lincoln.

Papa-Grimaldi, A. (1996) "Why Mathematical Solutions to Zeno's Paradoxes Miss the Point: Zeno's One and Many Relation and Parmenides' Problem" *Review of Metaphysics* 50(2): 299–314.

Papineau, D. (2013) *Philosophical Devices: Proofs, Probabilities, Pollibilities, and Sets* Oxford and New York: Oxford University Press.

Parker, R. (1983) *Miasma: Pollution and Purification in Early Greek Religion*. Oxford: Oxford University Press.

Pearson, N.J., J. Sarangi, and R. Fay (1996) "Evidence-Based Medicine" *British Medical Journal* 312: 380.

Peck, A. L. and E. S. Forster, tr. (1968) *Parts of Animals, Movement of Animals, Progression of Animals*. Cambridge, MA: Harvard University Press.Pellegrin, P. (1982) *La Classification des animaux chez Aristote: statut de la biologie et unité de l'aristotélisme*. Paris: Les Belles Lettrès.

_____. (1985) "Aristotle: A Zoology without Species" in *Aristotle on Nature and Living Things: Philosophical and Historical Essays in Honor of David M. Balme's Seventieth Birthday*, Gotthelf, A., ed. (pp. 95–115) Pittsburgh, PA: Mathesis Press.

Plato (1900–1907) *Platonis Opera*. 5 vols. J. Burnet, ed. Oxford: Clarendon Press.

Platt, A. (1921) "Aristotle on the Heart" in *Studies in the History and Method of Science*, C. Singer, ed. (pp. 520–532) Oxford: Oxford University Press.

Pliny (1951) *Natural History*, edited and translated by W. H. S. Jones. 10 vol. Cambridge, MA: Harvard University Press (Loeb).

Plutarch, "Themistocles" in *Plutarch's Lives*, vol. 2. edited and translated by Bernadotte Perrin (1914) Cambridge, MA: Harvard University Press (Loeb), 1–92.

Pohlenz, M. (1938) *Hippocrates und die Begundung der wissenschaftlichen Medizin*. Walter de Gruyter: Berlin.

Popper, K. (1959 [1934]) *The Logic of Scientific Discovery*. London: Hutchinson.

_____. (1972) *Objective Knowledge*. Oxford: Oxford University Press.

Potter, P. ed. (1988–2012) *Hippocrates, selected treatises*,vols. 5, 6, 8, 9, 10. Cambridge, MA: Harvard University Press (Loeb).

Praxagoras, (1958 [4th century BC]) *The Fragments of Praxagoras of Cos and His School*, edited and translated by F. Steckerl. Leiden: Brill.

Preus, A. (1970) "Science and Philosophy in Aristotle's *Generation of Animals*" *Journal of the History of Biology* 3: 1–5.

_____. (1975) *Science and Philosophy in Aristotle's Biological Works*. Hildesheim: Olms.

_____. (1977) "Galen's Criticism of Aristotle's Conception Theory" *Journal of the History of Biology* 10: 65–85.

Putnam, H. (1974) "The 'Corroboration' of Theories" in *The Philosophy of Karl Popper*, P. A. Schilpp, ed. The Library of Living Philosophers, vol. Xiv. LaSalle, IL: Open Court.

Quek, D.K. (1990) "Torsade de Pointes and Syncopal Attacks in a 26-Year-Old Woman with Congenital Complete Heart Block and Prolonged QT Interval" *Singapore Medical Journal* 31(2): 185–188.

Quine, W.V. (1987) *Quiddities: An Intermittently Philosophical Dictionary*. Cambridge, MA: Harvard University Press.

Ragep, F. J. (1990) "Duhem: The Arabs and the History of Cosmology" *Synthese* 83(2): 201–214.

Resnik, M. (1997) *Mathematics as a Science of Patterns*. Oxford: Oxford University Press.

Roberts, S. S. (2008) "A Women's Disease" *Diabetes Forecast*. 61(2): 48–49.

Robjat, D. (2012) "The Earthly Realism of Plato's Metaphysics or: What Shall We Do with Iris Murdoch?" *Philosophical Investigations* 35(1): 43–67.

Romm, J. (2012) *Ghost on the Throne: The Death of Alexander the Great and the Bloody Fight for His Empire*. New York: Vintage.

Rosenberg, A. (1985) *The Structure of Biological Science*. Cambridge: Cambridge University Press.

Rosenkranz, S. (2012) "Radical Scepticism Without Epistemic Closure" *Philosophy and Phenomenological Research* 85(3): 692–718.

Ross, W. D. (1923) *Aristotle*. London: Methuen & Co. Ltd.

Roussel, M. (1983), "La Notion de Traction dans le *Corpus Hippocratique*: vers une Étude Globale," in *Formes de Pensée dans la Collection Hippocratique*, F. Lasserre and P. Mundry, eds. (pp. 423–426) Geneva: Librairie Droz.

Rüsche, F. (1930) *Blut, Leben und Seele*. Paderborn: Ferdinand Schöningh.

Ruse, M., ed. (1989) *Philosophy of Biology*. New York: Macmillan.

Ryle, G. (1949) *The Concept of Mind*. London: Hutchinson & Co.

Sackett, D., W. M. C. Rosenbert, J. A. Muir Gray, R. Brian Haynes, and W. Scott Richardson (1996) "Evidence-Based Medicine: What It Is and What It Isn't" *British Medical Journal* 312: 71–72.

Sankey, H. (2012) "Scepticism, Relativism and the Argument from the Criterion" *Studies in History and Philosophy of Science* 43: 182–190.

Scarborough, J. (1976) "Celsus on Human Vivisection at Ptolemaic Alexandria" *Clio Medica* 11: 25–38.

Schöner, H. (1964) *Das Viererschema in der antiken Humoralpathologie*. Wiesbaden: Steiner.

Sedley, D. (1982) "On Signs" in *Science and Speculationi* J. Barnes, et al. ed. (pp. 239–266) Cambridge: Cambridge University Press.

Sextus Empiricus (1933) *Outlines of Pyrrhonism*, translated by R. G. Bury. Cambridge, MA: Harvard University Press.

Shakespeare, W. (1969) "The Winter's Tale" in *The Complete Pelican Shakespeare*, Alfred Harbage, ed. Baltimore, MD: Penguin, 1334–1368.

Sharma, R. (2012) "Diocles of Carystus on Scientific Explanation" *The Classical Quarterly* 62(2): 582–602.

Shaw, J. R. (1977) "A Note on the Anatomical and Philosophical Claims of Diogenes of Apollonia" *Apeiron* 11:5 3–57.

Sherry, D. (1988) "Zeno's Metrical Paradox Revisited" *Philosophy of Science* 55: 58–73.

Smith, C. U. M. (2013) "Cardiocentric Neurophysiology: The Persistence of a Delusion" *Journal of the History of Neurosciences* 22(1): 6–13.

Smith, W. (1990) *Hippocrates. Pseudepigraphic Writings. Letters, Embassy, Speech from the Alter, Decree*. Leiden: Brill.

Solmsen, F. (1961) "Greek Philosophy and the Discovery of the Nerves" *Museum Helveticum* 18: 185–188.

Sophoclis Fabulae (1924) Ed. A. C. Pearson. Oxford: Oxford University Press.

Sophocles "Philoctetes" (1957) Tr. D. Grene. *Sophocles II*. Chicago, IL: University of Chicago Press.

Soranus, *Die Überlieferung der Gynäkologie des Soranos von Ephesos*. (1911) Ed. J. Ilberg. 4th ed. Liepzig: Teubner.

Soranus, *Gynecology*. (1956)Tr. O. Temkin. Baltimore, MD: Johns Hopkins University Press.

Sorbji, R. (1980) *Necessity, Cause, and Blame: Perspectives on Aristotle's Theory*. Ithaca, NY: Cornell University Press.

Steckerl, F. (1958) *The Fragments of Praxagoras of Cos and His School*. Leiden: Brill.

Sterelny, K. and P. Kitcher (1988) "The Return of the Gene" *Journal of Philosophy* 7: 339–361.

Strang, C. (1961) "The Perception of Heat" *Proceedings of the Aristotelian Society*. 41: 239–252.

Striker, G. (1986) "Antipater and the Art of Living" in M. Schofield and G. Striker, eds. *The Norms of Nature*. Cambridge: Cambridge University Press.

Struck, P. (2004) *Birth of the Symbol*. Princeton, NJ: Princeton University Press.

Szabó, A. (1978) *The Beginnings of Greek Mathematics*. Synthese Historical Library, vol. 17. Dordrecht: D. Reidel.

Tambiah, S. J. (1985) "The Form and Meaning of Magical Acts: A Point of View" in *Culture, Thought, and Social Action*. Cambridge, MA: Harvard University Press.

Tecusan, M. (2004) *The Fragments of the Methodists: Methodism Outside Soranus*. Leiden: Brill, 2004.

Tertullian (rpt. 2007) *De Anima*. J. H. Waszink, ed. Hildesheim: Olms.

Thivel, A. (1975) "Le dévin dans la collection hippocratique" in *La collection hippocratique et son rôle dans l'histoire de la medicine*, L. Bourgey and J. Jouanna, eds. (pp. 57–76) Leiden: E. J. Brill.

Thompson, G. (1941, rpt. 1972) discusses this in his book *Aeschylus and Athens*. New York: Haskell House.

Thrämer, E. (1894) "Asklepios (2)" *Pauly-Wissowas Realencyclopädia dell'arte antica classica e orientale* 2 cols. (pp. 1642–1697) Stuttgart: J. B. Metzler.

_____. (1914) "Health and Gods of Healing (Greek)" *Encyclopedia of Religion and Ethics* 6: 540–553.

Tieleman, T. (1998) "Plotinus on the Seat of the Soul: Reverberations of Galen and Alexander in Enn. IV, 3, 27ESS, 23" *Phronesis* 43(4): 306–325.

_____. (2008) "Methodology" in *The Cambridge Companion to Galen*, R. J. Hankinson, ed. (pp. 49–65) Cambridge: Cambridge University Press.

Tsakmakis, A. and M. Tamiolaki, eds. (2013) *Thucydides Between History and Literature*. Boston and Berlin: Walter de Gruyter.

Valtonen, Mauri and Hannu Karttunen, (2006) *The Three Body Problem*. Cambridge: Cambridge University Press.

Van der Eijk, P. J. (1990) "The Theology of the Hippocratic Treatise *On the Sacred Disease*" *Apeiron* 23: 87–119.

_____. (1999) "Historical Awareness, Historiography and Doxography in Greek and Roman Medicine" in *Ancient Histories of Medicine: Essays in Medical Doxography and Historiography in Classical Antiquity*, P. J. Van der Eijk, ed. (pp. 1–33) Leiden; Boston: Brill, 1999.

_____. Ed. (2000) *Diocles of Carystus: A Collection of the Fragments with Translation and Commentary*, volume 1. Leiden: Brill.

_____. Ed. (2001) *Diocles of Carystus: A Collection of the Fragments with Translation and Commentary*, volume 2. Leiden: Brill.

_____. (2001) "The Systematic Status of Therapy in the Hippocratic Corpus and in the Work of Diocles of Carystus" *Aspetti della terapia nel Corpus Hippocraticum : Atti del IXe Colloque International Hippocratique* (Pisa, 25–29 Settembre 1996) in *Medical History* 45(1): 389–404.

_____. (2005a) *Medicine and Philosophy in Classical Antiquity*. Cambridge: Cambridge University Press.

_____. Ed. (2005b) *Hippocrates in Context: Papers Read at the XIth International Hippocrates Colloquium University of Newcastle upon Tyne 27–31 August 2002*. Leiden: Brill.

______. (2008) "The Role of Medicine in the Formation of Early Greek Thought" in *The Oxford Handbook of Pre-Socratic Philosophy*, P. Curd and D. W. Graham, eds. (pp. 385–412) Oxford: Oxford University Press.

_____. Ed. (2012) Neil Allies, tr. *Greek Medicine from Hippocrates to Galen.* (pp. 3–20) Leiden: Brill.

van der Eijk, P., H. F. J. Horstmanshoff, and P. H. Schrijvers, eds. (1995) *Ancient Medicine in Its Socio-Cultural Context: Papers Read at the Congress Held at Leiden University, 13–15 April, 1992*, vol. 2. Amsterdam: Rodopi.

Van Fraassem. B. C. (1976) "To Save the Phenomena" *The Journal of Philosophy* 73: 623–632.

Vassale, C., S. Maffei, et al. (2008) "Gender-Related Differences in Oxidative Stress Levels in Elderly Patients with Coronary Artery Disease" *Fertility and Sterility* 89(3): 608–613.

Vegetti, M. (1995) "L'Épistémologie d'Érasistrate et la Technologie Hellénistique" *Ancient Medicine in Its Socio-Cultural Context; Papers Read at the Congress Held at Leiden University, 13–15 April 1992*. 28: 461–472.

Vermeir, K. (2005) "The Magic of the Magic Lantern (1660–1700): On Analogical Demonstration and the Visualization of the Invisible" *British Journal for the Philosophy of Science* 38(2137): 127–159.

Vesalius, A. (1552) *De humani corporis fabrica*. Lugduni: Joan Tomaeium.

Vlastos, G. (1953) "Isonomia" *American Journal of Philology* 74: 337–366.

von Staden, H., ed. (1989) *The Art of Medicine in Early Alexandria.* Cambridge: Cambridge University Press.

_____. (1992) "The Discovery of the Body: Human Dissection and Its Cultural Contexts in Ancient Greece" *The Yale Journal of Biology and Medicine* 68: 223–241.

Wasserstein, A. (1962) "Greek Scientific Thought" *Proceedings of the Cambridge Philosophical Society*. N.s. 8: 51–63.

Webb, O., MD. (2012) *How We Do Harm: A Doctor Breaks the Rules about Being Sick in America.* New York: St. Martin's Press.

Webster, C. (2005) *From Paracelsus to Newton: Magic and the Making of Modern Science*. New York: Dover.

Wellmann, M. (1901) *Die Fragmente der sikekischen Ärtzte Akron, Philiston und des Diokles von Karystos*. Berlin: Weidmann.

_____. (1922) "Der Verfasser des Anonymus Londinesis" *Hermes* 57: 396–430.

Wilcox, M. (2012) "Improved Traditional Medicines in Mali" *Journal of Alternative and Complementary Medicine* 18(3): 212–220.

Wilde, L. W. (2000) *On the Trail of the Women Warriors: The Amazons in Myth and History*. New York: Macmillan.

Williams, G. (1989). "Adaptation and Natural Selection" in *Philosophy of Biology*, M. Ruse, ed. (pp. 182–185) New York: Macmillan.

Wimsatt, W. (2006) "Reductionism and Its Heuristics: Making Methodological Reductionism Honest" *Synthese* 151(3): 445–475.

Winter, M. (1997) "Aristotle, hos epi to polu Relations and a Demonstrative Science of Ethics" *Phronesis* 42(2) (1997): 163–189.

Woudenberg, R. V. (2007) "Doxastisch Antirealisme" *Algemeen Nederlands Tijdschrift voor Wijsbegeerte* 99(3): 188–202.

Wright, T. (2013) *Circulation: William Harvey's Revolutionary Idea*. London: Vintage.

Yahuda, A. S. (1947) "Medical and Anatomical Terms in the Pentateuch in Light of Egyptian Medical Papyri" *Journal of the History of Medicine and Allied Sciences* 2: 549–574.

Names and Subjects Index

Adkins, A. 18n7n9
aer, 3, 34–36
Aeschines, 18n6
Aeschylus, 9, 19n23, 48n72, 33, 40
Aëtius, 10, 22n70
air, 10; air_1 117, 122, air_2 117, 132n54 equivalent to $pneuma_1$, *see also* *aer, pneuma*
Aird, W. 132n51, 133n75, 134n82
aither, 34–36
Agamemnon, 9
agathos, 2
Alcmaeon of Croton, 9, 10, 12, 17, 33, 59, 61 *isonomia,* 15; separates sense from reason, 11
Alexander of Macedon, 102n71
als-ob discourse, 4, 19n25
Amyntas III of Macedon, 49
anastomosis, 117, 121
anatomy, 94, *see also* dissection, dissection, simian, vivisection
Anaxagoras, 14, 21n34, 40, 41
Anaxarchus, 18n12
Anaximander, 14, 23n89
Anaximenes, 14, 23n89
animals, blooded and non-blooded 55, 58, 71n33
animal spirits 118 (*see also* $pneuma_1$)
Anonymous of Brussels, 99n8, n13
Anonymous Parisian, 46n42, 100n27
Anonymus Londinensis, 18n1, 25, 35, 36, 37, 38, 43n2, n6, 79, 100n33, 134n86
anti-realism, *see* epistemology, anti-realism
aorta, 74n72, 79, 80
arche, 57, 62, 70n19, 72n38, 82
arete, 2, 45n29, right side more *arête* 60
Aristophanes, *Lysistrata* 48n76
Aristotle, 2, 4, 7, 8, 9; 12, 14, 15, 16, 18n. 5, 19n16, 22n51, n57, n58, 23n94, 25, 26, 32, 35, 44n.14, 45n23, 46n42, 47n56, ch. 3, 77, 78, 80, 83, 92, 99n2, n7, n18, 111, 128, 130n20, 132n48; as biologist 49, brain as cooling unit 130n11, conception theory 40, 41, 45n21, 47n50, 105n114, explanatory principles, 90, 112, 97, 105n113, four causes 52, 112, heart as the source of reason 11, origin of nerves 82, presentation of science 88, 90, 112, redundant explanation 53, synthetic/systematic philosophy 110, *techne* and *methodos* 53, theory of three souls 10, 56
Arrian, 102n71
art [of medicine, *techne*], 31, 32, 45n29
arteries, 10, 11, 32, 80, 81, 112, 117; carrying blood and *pneuma* 85, carrying only *pneuma*, 86, coatings 85, 119, emanated from right side of heart 120, pores 120, pulse 122, 123–125; thick 79, *see also phlebes*
asklepieion, 3, 7, 85
Asclepiades, 41, 81, 108n147
Asclepius, 3
Ashmore, M., 134n94
Asmis, E., 108n142
astronomy 23n93, 49
ataraxia, see quietude
atomists, 53
Atreides, 7
attraction (*holke*), 114
Augustine, 73n66, 106n122
autopsy, 84
Ayala, F., 44n15
Aziz, M., 132n51

Babb, J., 107n137
Baker, M.A., 73n65
Balaguer, M., 19n24
balance and health, 4, 15, 32
Balme, D., 44n12, n14, 70n14, 71n25, 73n66, 74n91, 75n117
Baltzly, D., 24n100
Bardong, K., 100n31
Barnes, J., 44n14, 70n17, 108n159
Batens, D., 19n25
Beardslee, D., 18n13
Belkin, G., 72n38
Berkeley, G., 105n116
Berryman, S., 44n19, n20, 101n49, 130n14, 131n37
bilateral symmetry, 16, 78, 86, 90, 103n86
bile, 41; and *cacochymia* 135n98, as a humor 114, yellow and black 79, 118
biomedicine, theory approach 26, 69, 98, 106n134, (*see also* Dogmatism), discovery and explanation 55, and *epi to polu*, 28, 32, 49, 75n119, 92, evidence-based approach 26, 43n9, 45n22, 47n55, 77, 92, 106n134, 110, 114, 132n41, general statements 85, laws, 42, 47n55, table integrating categories 92; *see also* empiricism, *see* scientific explanation, in the context of modern theory 103n101
Birch, T. 134n78
Blondlot, R., 134n94
blood, 12, 41, 42; and air 34–36, 74n94, Aristotle's theory of classification 55–58, blood as humor 114, blood to the lungs 130n18, and the brain 61, as the cause of disease 38, coagulation 59, color 59, 71n33, consistency 59, 60, creation of 62–63, 65, conception 67, 97, and courage 7, 8, and female menses 14, 97, 98, and *ichor* 2, 8, 9, 18n12, 32, 33, 56, 77, insects 73n62, male v. female 75n101, n117, and nourishment 56, 71n33, 134n90, origins 21n50, and *pneuma* 3, 56, 87, 117–119, and *psuche* 10, 29, 56, and *puon* 38, and reason 60, 75n98, and skin 12, and *thumos* 7, too much or too little, 15, 39, 97, 126, warm fluid 37, *see also ichor, katamenia*
blood vessels, 2, 10, 16, 39, 42, 57, 58, 59, 62, 66, 77, distinguishing arteries from veins, 79, 80, and pulse 58
body parts, homeomerous, 12, 58, 64, 72n38, anomeromerous 12, 58, 64, 72n38; *see also* specific organs
Booth, N.B. 22n75
Bos, A., 73n66
Boudon-Millot, V., 44n11
Bourgey, L. 23n90
Boyd, R. 23n81
Boylan, M. 18, n.1, 21n, 33, n35, n38, n50, 22n51, n54, n57, 67, n68, 23n82, n85, n86, n87, n90, 24n100, 44n12, n14, 45n22, 47n50, n52, n55, n57, n60, 48n72, 69n2, n3, n5, n6, 70n14, 74n.81, n92, 75n97, 101n52, n66, n69, 105n112, 105n117, 106n124, n126, 107n138, n139, n152, 131n22, n27, 131n35, 132n44, 132n47, n49–50, 132n55, 134n93
Boyle, R. 44n19
Brain, P., 135n97, n102
brain, 34, 35, 58; Aristotle's argument 61, as the seat of reason 11, 46n42, 62, 129, and blood 61, 62, damage 135n100, major organ 119, 129, origin of the spinal cord 112, and *pneuma* 118, source of nerves 86, two lobes 78
Brock, S., 19n. 24
Brown, K., 21n35, 22n68, 23n82, 45n22, 47n55, n60, 132n55, 134n93
Buchowski, M., 20n28

cacochymia, 135n98
Carastro, M., 2, 19n15, n.16, n18
careful scrutiny approach, *see* epistemology, careful scrutiny maxim
Carnap, R., 134n86
Carneades, 90
causation in science, 5; divine accounts 6, 8, 13, 17; 20n28, 45n32; material accounts, 6, 8, 13, 17, 20n28, 45n32, 52, 71n26, 53,

72n38, 80, 86, 112, 116; *see also* scientific explanation, ME.
Celsus 82, 93, 98, 100n31, 101n62, n64, 108n159; the evident and the hidden 82–83
chance, 3, 14, 76n123
Chang, H,. 99n4
Chaucer, G., 20n28
Churchland, P., 106n.130
Chrysippus, 82
Cicero, 21n36, 23n80
Circle of Willis, 73n65
circulation, 73n62
Cleophanes, 41
Clifford, W., 105n112
clinical approach to medicine/practice, 16, 37, 38, 77, 92, 93, 94, 110, 132n41; *see also* empiricist school
Cnideans, *see* Hippocratic writers, Cnideans
cold (as in hot, cold, wet, dry); *see* contraries
Collins, D., 2, 18n13, 19n15, n18, n43
conception theory, 39–43, 65–69, 97; blood 67, dual seed 40, 41, 75n101, epigenesist 41, 67, gender expression 66, 68, gender seed dynamics 68, 75n101, male single seed 40, 41, 65–69, monthly cycle 42, pregnant women 42, pangenesis 41, 67, women 40
concoction (*pepsis*), 60, 63, 64, 65, 66, 67, 80
constitutions (*katastasis*), 30
contingency (logical operator ◊), 26, 30, 31, 32, 36, 37, 38, 39, 42, 45n21, n24, 46n43, 50, 61, 69, 77, 90, 92, 93, 98, 113, 122, 125, 126, 128, 135n96
contraries, 12, 27, 31, 37, 54, 57, 59, 60, 63, 64, 66, 73n59, n65, 75n103, 109, 111, 116, 117, 118
Cooper, J., 71n25, 75n119
Copernicus, 125
corpses, prohibition on touching 83
Cos writers, *see* Hippocratic Writers, Cos
Cosans, C., 130n14, 132n47, 133n77
crisis (*ekrithe*), 30

Dancy, J., 45n21
Darenberg, C., 18n3, 131n25
Darwin, C., 125
Davidhoff, F., 106n134
Dean-Jones, L., 23n93
death, 36, 58, 73n.64, and reason 62
deductive-nomological method, 27, 42, 103n101, 114; *see also* scientific explanation
definitions, 50, 55, of humans 56
Deichgräber, K., 106n128, n132
Democritus, 19n. 16
DeLacy, P., 130n19, 131n36
Demont, P., 18n4, n11, 32, 33, 39, 45n23, n31, 46n37, n39, 41, 46n45, 48n69, n70, 75n98, 108n155
Descartes, 22n55, 44n19, 47n50, 89, 104n109, 118, 133n56; and eristic philosophy 110
Dexippus, 38
diagnosis, 31, 38, via pulse 82, 86, 116, 128, via *historia* 115
Dicker, G., 105n110
Dicks, D., 23n93
Diels, H., 25, 43n3, 47n49, 100n33
Dieuches, 100n28
digestion, 38, 58, 64, 65, 80, 81
Diocles of Carystus, 77, 78, 81, 82, 92, 98n1, 99n7, n8, n18, n19, 100n22, n26
Diogenes Laertius, 9, 11, 14, 22n60, n71
Diogenes of Appollonia, 16
disease, 6, 37, 78, 90, 116, 132n42, n44, common and particular 35, focus of Methodists 96, imbalance of humors 114, three-day cycle to recovery 96, sacred 21n40, 32
dissection, 81, simian 111, 130n15
divinity, 3, 7, 13, 23n89; *see also* $phusis_2$
Dobson, M., 135n111
Dodds, E.R., 2, 7, 21n46, 22n52, 23n84
Dogmatists, school of, 6, 36, 49, 61, 77, 78–88, 91, 92, 96, 98, 107n135, 109, 125
Doney, W., 105n110
"Do no harm," 95
dry (as in hot, cold, wet, dry); *see* contraries
Dryer, J.L.E., 23n93, 69n4, 74n71
dual seed theories, *see* conception theory, dual seed
Ducatillon, J., 46n32

Duminil, Marie-Paule, 18n7
Dunamis, 29, 35, 36, 38, 40, 41, 42, 47n52, 57, 86, 90, 118, 119, 122

Edelstein, L., 2, 7, 9, 19n15, n17, 21n33, n47, 22n61, 43n1, 106n128, 108n147, n148, n154, 134n92
Eidos; see species
Einstein, stand-alone systematic scientist (natural philosopher) 110, 125
Egypt, 84
Ellinger, T. 48n79
embryo, 57, 58
emergent scientific explanation, *see* explanation-type β
emotion, *see thumos*
Empedocles, 5, 9, 12, 17, 35, 37, 41, 47n56, 72n57, 78; four roots 11, 13, 14, 27, 37, 111
Empiricists, school of (and Empirics), 5, 6, 21n.33, 36, 42, 43n9, 77, 88, 89, 93–96, 110, 125; and behaviorism 93, tripod: autopsia, historia, similarity of connection, 93
Energia, 47n52
Entelechia, 47n52
Epicurus, 99n3
epidemic 46n45
epigenesis, *see* conception theory, epigenesis
epistemology, realism, anti-realism, 3, 4, 5, 6, 20n27; appearance v. reality problem 83, careful scrutiny maxim 89, 90, 91, 92, 110, closure 89, certainty 90, selected faith maxim 90, 91, 92, skepticism, 3, 6, 50, 88–93; table interrelating categories 92, *see also* skepticism
episteme, 4, 50
Epi to polu; see 'for the most part'
Erasistratus, 35, 82, 83–88, 92, 93, 103n95–97, 117, 118, n99, 120, 122, 123, 124, 125, 126, 128, 134n83, 129n1, pulse experiment 87
eristic philosophers, *see* philosophy eristic
Eudemus, 85
Euripides, 48n73, 40
Eurymachus, 41
Euryphon of Cnidos, 46n34
Euthyphro, 5
evidence-based approach, *see* biomedicine evidence-based approach
explanandum, material, 3, 91, 109, 125
explanans, material, 3, 20n28, 28, 53, 71n26, 72n38, 80, 86, 92, 130n20
explanation-type α, 28, 34, 35, 36, 37, 42, 47n51, 53, 54, 60, 66, 98, 106n125, 109, 112, 114, 119, 126, 128, 131n37
explanation-type β, 28, 37, 60, 92, 98, 106n125, 126, 128

female seed, *see katamenia*
Ferejohn, M., 44n18
fevers (*puretoi*), 35, 95, 126, 127; hunger cures 80, venesection cure 126
Flemming, R. 129n2, 131n35
Fodor, J., 106n130
Forrester, J.M., 104n106
for the most part, 32, 49, 75n119, 92, 113
Francis, S., 69n2
Fraser, P.M., 102n83
Frede, M., 45n23, 99n5, 106n128, 112, 134n92
Frederich, C.J., 22n69, 46n34
front preferred to back, 72n54
Fuchs, R., 100n45
Furley, D. 22n75, 131n, 32, 134n80
Furth, M., 75n119

Galen, 14, 18n5, 22n51, 57, 31, 44n11, 48n78, 61, 70n19, 72n49, 73n65, 74n80, 75n112, n113, 77, 78, 82, 87, 88, 95, 96, 97, 98, 99n3, n7, n10, n.23, 100n28, 101n50, n56, n58, 102n84, 103n85, n93, 107n140, 108n143–145, 108n159, 115 ch. 5; brain responsible for thinking 130n16, bias against fruit 131n39, disease as imbalance of humors 114, Cos/Cnidean (Sicilian) distinction 115, origin of arteries 112, pulse and bile 133n68, respiration experiment 122–23, simian dissection 111, spinal cord begins at the brain 112, studied dissection in

Alexandria 130n15, too much phlegm 114, using Aristotle 112
Galileo, 129n7, stand-alone systematic scientist (natural philosopher) 110
Garafalo, I., 103n100
Gardner, M., 134n94
genetic order, 51
genos; see genus
genus 26, 51
Gewirth, A., 105n110
Gill, C., 130n12
Gimbel, S., 103n101
gods, 3, 7, 8, 18n14, 23n89, 70n19, 85; *see also* divinity and *phusis*$_2$
Gomperz, T., 21n36
gonads, 42
good, 53; *see also* teleology
Goodman, N., 24n102, 134n86
Gordon, R., 2, 18n13, 19n15, n16, n18, 21n39
Gotthelf, A., 69n7, 70n14, 73n66
Grams, L., 47n59
Graz, B. 107n137
Grene, M., 69n2
Grice, P., 44n10
Gourevitch. D., 43n2
Gundert, B., 38, 47n54, 48n65
Gutherie, W.K.C., 9

Hacking, I. 44n17, 104n102, n107
haima, see blood
Hall, T., 23n78
Hankinson, R.J., 2, 18n10, n13, 19n15, n18, 21n38, n42, 22n58, 23n92, 45n23, 105n118, 105n128, 108n159, 130n9, 131n25, n26, 134n92
Harris, C.R.S., 9, 14, 22n65, n69, 23n83, 46n33, 59, 62, 72n49, 74n72, 99 n.6, n14, n17, 100n46, 103n88, n94, 104n103, 106n133, 131n28, 132n46, 133n69, n70, 71n74
Harvey, W., 84, 88, 104n106, 122, 132n51, 133n69
health, 37, 47n57, 90, 132n44
heart, 7, 8, 58, 64, 74n72, 77, 112, 120; as the seat of reason 11, 78, as *arche* 57, 64, as knot of blood vessels 62, motions of [pulse] 64, 86, 119–123, creates blood 65, chambers 78, 86, 100n46, contractions 121, expelling *pneuma* 121, not a pump but a periodic release of *pneuma,* origin of nerves 82, 86, systole 94, source of power, major organ 119, 129, *see also* pulse
heavenly bodies, 15
Heinimann, F., 18n13
Hegel, G.W. F., stand-alone systematic philosophy 110
Held, B., 19n24
Hempel, C., 134n86
Henry, D., 69n7, 70n12, 76n119
hepatic vein, 112
Heraclides of Tarentum, 96
Heraclitus of Ephesus, 23n89, 89, 92
Herodotus, 18n2, 102n73, n74
Herophilus of Chalcedon, 33, 73n65, 79, 82, 83–88, 92, 93, 94, 100n45, 102n79, 103n86, 103n87, n88–92, 119, 120, 125, 128, 133n70
heuristic discourse, 4
Hippocrates (Hippocratic writers), 1, 2, 9, 10, 14, 15, 18n1, 18n5, 22n51, n59, 23n88, 25, 34, 35, 39, 40, 41, 42, 44n11, 60, 67, 75n113, 77, 79, 82, 107n136, 109, 111, 113, 120, 127, 128; Cos writers 11, 12, 15, 23n90, 25, 31, 34, 35, 113; Sicilian/Cnidean writers 11, 25, 31, 46n43, 61, 78, 82, 113, 114, 115
Hoffman, R. 135n99
Homer (Homeric writers), 1, 7, 15, 17, 18n2, 22n53, 46n36, 72n49
Hong, H., 135n111
Horowitz, M. 23n87
Horstmanshoff, H. F. J., 99n2
hot (as in hot, cold, wet, and dry); *see* contraries
Hudnall, S., 135n99
Hudson, R., 24n100
Hull, D., 44n15
Hume, D., 19n26, 89, 104n108, 107n141, 132n45; and eristic philosophy 110
humor theory, 12, 15, 32, 38, 94, 107n135, 114, 116, 118, 132n40
Hüttemann, A., 105n110
Huxley, T., 72n49
Hwa, C., 133n75, 134n82
hypothesis [as a method], 32

Iamblichus, 73n66
Ichor, 1, 2, 8, 9, 10, 13, 14, 17, 18n11, n12, 33, 56, 57, 58, 59, 65; 72n52, 75n98, 85, 97, 98; *see also puon*
identity, principle of 55; *see also* similarity
Ilberg, J., 135n101
intelligence, men v. women, 73n59
Isonomia, 15

James, H., 46n47
James, W., 105n117
Jenkins, M., 129n4
Joly, R., 48n71
Jones, W., 43n5
Jori, A., 72n56
Jouanna, J., 18n4, n11, 23n90, 33, 45n23, 46n37, n39, n41, 47n54, 75n98, 102n74, n80, 108n155

Kahn, C., 18n13
Kant, I., 6, 20n27. 21n41, 24n102, 45n21
Karshren, E., 20n27
Karttunen, H., 106n123
Katamenia, 14, 23n86, 40, 42, 63, 65, 66, 67
Katharsis, 85
Kenos, 35, 86
Kiblansky, R.E., 131n38
kidneys, 41, 42
Kitcher, P., 44n15
Kleos, 2
Kuhn, T., 19n25, 74n71
Kullmann, W., 71n25
Kuriyama, S., 135n111

Lakatos, I., 44n13
Lange, M., 23n93
Lattimore, R., 18n8
Lefkowitz, M., 48n75
Leibnitz, G., principle of sufficient reason, 5, 17, 35, stand-alone systematic approach 110
Leith, D., 108n150
Lesky, E., 48n73
Leucippis 19n16, 41
Levin S., 74n67
Limet, R., 132n51
liver, 58, 80, 81, major organ 119, 129
Lloyd, G.E.R., 2, 15, 18n10, 13, 19n15, n17, n18, 23n84, n92, n99, 35, 43n1, 45n23, 46n44, 47n48, 101n64, 102n82, 129n2, 131n33, 132n40
Lock, M., 72n38
logical empiricism, 27
logical order, 51
logikoi, school of; *see* Dogmatists
Logino, H., 19 n.24, 135n95
Lones, T.E., 73n64
Long, A.A., 21n33
Longrigg, J., 99n2, 103n87
Lonie, I., 23n86, 46n43, 47n34, 48n71, n79
Lorenz, K., 103n101, n102
lungs, 39, 53, 81, 113, 117, 118, 120, 129, 130n18; pulmonary circulation 122

ME; *see* Scientific explanation, ME
magic, 2, 3, 4, 5, 6, 7, 8, 14, 15, 17, 18n13, 20n27, n28, n29, 23n89, 35, 84, 92, 102n75, 108n156
Magnus, D., 19n25
male single seed theory; *see* conception theory, male single seed
Manetti, D., 25, 43n7
Mani, N., 130n19
Mares, E. 19n24
Marganne, M., 135n109
Martini, D., 133n60
material, *see* causation in science, material
mathematics, 15, 49, 62, modern laws of probability 88,
Matthen, Mohan, 107, n134
Mattern, S., 135n112
Mayhew, R., 75n115, n116
McClure, L., 48n75
mechanical accounts, 27, 86, 111, 126, 130n14
Menon, 25
menses, *see katamenia*
Methodists, school of, 6, 36, 49, 77, 88, 89, 96–98, 108n154, 110, 125, 132n42, could be learned in 6 months 96, popular in Roman Empire 96, three-day cycle to recovery 96
Michael of Ephesus, 74n91
Mill, J.S., 103n102
Milton, J., 46n39, n41
Mnesitheus, 100n28
Modal Continuum of Explanatory Structure within Biomedicine, 28

modal operators; *see* necessity, contingency
More, H., 44n19, 47n50
Morrison, K., 19n24
Morsink, J., 75n119
Mucz M., 107, n137
Mueller, I., 45n27, 69n4
Mullins, W., 134n79

nature, abhors a vacuum, 17, 35, 86, against nature 36, 37, 38, 40, 76n123, according to nature 37, 40, 69, 134n83, capital 'N' 130n13, does nothing in vain, 35, 55, 83, 90, 111, 121, does nothing haphazardly 70n19, male and female natures 42, 67–69, *see phusis*
Nawer, T., 129n4
necessity (logical operator □), 5, 19n26, 26, 30, 32, 36, 37, 38, 39, 42, 45n21, 49, 50, 55, 61, 91, 92, 106n125, 109, 113, 120, 121, 122, 125; two sorts of Aristotelian necessity 54, 55, 69, 77, 126, 128, 135n96
Needham, J., 48n72
nerve (*neuron*), 33, motor, sensory, and cranial, 86; *see also pneuma*
Newton, I. 27, stand-alone systematic scientist (natural philosopher) 110, 125
Newtonian physics, 91
Nicarchus, 79, 81
Nikel, D., 100n33
nomos, 18n13, 90
nourishment, *see trophe*
nous, 21n34
Novotny, J., 48n64
Nussbaum, M., 23n85
nutrition, 55, 56, 119
Nutton, V., 2, 7, 19 n.15, n20, n22, 21n44, n46, 24n101, 45n23, n32, 46n46, 69n1, 84, 98n1, 99n7, n9, n11, n14, n18, 100 n, 21, n26, n28, n39, 102n72, n83, 104n106, 106n128, 108n146, n147, n149, 129n1, 130n10, 131n38, 131n39, 132n40, 133n57

O'Brien, D., 22n75
Ogle, W., 72n49
Olson, S., 18n9
ontology, 11, 91,109, 111; *see also* phusis and nature

Paley, W., 44n19
pangenesis, *see* conception theory, pangenesis
Papa-Grimaldi, A., 105n114
Papineau, D., 44n10
Parker, R., 21n44
Parmenides, 41, 89, 92
particularists, 45n21
Pearson, N.J., 106n134
Peck, A.L., 73n60
Pellegrin, P., 44n12, 69n7, 70n17
Peloponnesian war, Athenian plague 46n45
pineal gland, 86, 118
Pericles, 46n45
Pharmakon 21n43, 85, 102n73
Philiston, 12, 79
Philolaos, 18n12
Philolaus of Tarentum, 10
philosophical approach to biomedicine, 15, 77, 92; *see also* Dogmatism
philosophy, systematic v. eristic, 73n59, 109, 110, 129, 129n5
phlebes, 2, 10, 23n95–97, 32, 34,79; *see also* blood vessels, veins
phlegm, 34, 41, 78, 79, 114; as a humor 114
phusis, 4, 106n125, 132n43; *kata* and *para; see* nature, according to, against
phusa, 34–36
phusis$_1$, 2, def. 3, 4, 5, 6, 7, 11, 12, 13, 15, 16, 17, 25, 27, 28, 30, 32, 33, 34, 35, 37, 39, 50, 51, 53, 54, 55, 57, 58, 60, 67, 69, 77, 78, 79, 91, 92, 96, 102n75, 103n86, 109, 114, 116, 119, 122, 126
phusis$_2$, 2, def. 3, 4, 5, 6, 7, 10, 11, 12, 13, 14, 15, 16, 17, 19n26, 20n27, 21n34, 25, 28, 33, 35, 36, 37, 39, 45n32, 46n45, 50, 51, 53, 58, 60, 62, 67, 68, 69, 72n44, 84, 85, 86, 91, 92, 95, 96, 97, 103n86, 108n156, 120, 121, 122, 130n13, 131n35
phusis$_3$, def. 3, 4, 5, 6, 7, 11, 13, 14, 15, 17, 19n26, 20n27, n28, 25, 28, 32, 36, 37, 39, 42, 60, 70n10, 92, 98, 110, 126, 131n37

Phylotimus, 79, 80, 100n28, 101 n, 50
physiology, 94, *see also* dissection, vivisection
Plato, 12, 18n5, 22n52, 25, 45n23, 47n56, 48n73, 49, 79, 133n61; divided line, 32, 45n28, and eristic philosophy 110, and the heart 62, 74n68, 105n113, remarks on science 37, 40, 45n21, 46n38, 62, 63, 72n52, 74n67, n70, 99n3, 100n24, 111, soul and body 130n9, theory of Forms 95, 110, 105n113, tripartite soul 111, 128, truth theory129n2
Platt, A., 72n49
Pliny, 4, 20n29, 102n73
Plistonicus, 79, 100n28, n37, 42
Plutarch, 4, 20n29
pneuma, 3, 10, 12, 14, 34–36, 54, 59, 64, 73n66, 74n94, 78, 79, 80, 86, 113, 119, 126, and blood 87, equivalent to *aer*$_2$, *pneuma*$_2$ 118, nerve-like activity 81, 112, psychic 112, 118, 124, types: *pneuma*$_1$ 118, *pneuma*$_3$ 13, *see also* animal spirits
Pneumatists school, 99n6, 131n28, 132n46
Pohlenz, M., 43n1
Polybus, 16, 46n42
Popper, K., 44n13, 104n102, 134n86, n91, 135n95
Praxagoras of Cos, 11, 33, 77, 78, 79–82, 92, 99n7, 100n28, n29, n30, n32, n34–36, n38, n40–41, n44, n45, 101n47–48, n50–51, n55, n59, n61, 103n86, 125, 128
pre-formation, 117: *see also* conception theory, epigenesis
pre-Hippocratic Writers, 3, 7, 13, 33; *see also* pre-Socratic writers
pre-Socratic Writers, 3, 5, 9, 11, 14, 18n2, 23n89, 30, 45n23, 109
Preus, A., 23n87, 48n73, 73n65
principle of sufficient reason, *see* Leibnitz, G. principle of sufficient reason
prognosis, 31, 116
prolepsis, 47n53
Protagoras 45n29
psuche 10, 29, 56, 69, and body 57, 71n36, rational 69, *psuche*$_1$, 15, *psuche*$_2$, 15
pulmonary artery, vein, 85, 102n84, 120, 121
pulse, 81, 86, 87, 119–123; *see also* heart, pulse
puon, 33, 59
Putnam, H., 134n91
putrefaction, 59, 97, 126, 128
Pythagoras, 7, 9, 10

Quek, D.K., 44n16
quietude, 90
Quine, W., 44n10

Ragep, F., 23n93
realism, *see* Epistemology, realism
reason, 56, 57, 71n26
regimen, 94
reproduction, 13, 55, 66; *see also* conception theory
residue (*perittoma*), 37, 38, 39, 58, 63, 64, 126, 128
Resnik, M., 19n24
respiration 12, 54, 64, 81, skin breathing 78, 113, 118, 131n30
rete mirabile, 86
right side of the body over left, 60, 72n57, 109, 120
Roberts, S., 44n16
Robjant, D., 129n6
Romm, J., 102n71
Rosenberg, A., 44n13
Rosenkranz, S., 105n111
Ross, W.D., 9, 73n66
Roussel, M., 47n54
Royal Academy, 134n78
Rufus of Ephesus, 79, 102n84
Rüsche, F., 7, 10, 21n48, 22n66, 23n83
Ryle, G., 73n66, 129n6

Sackett, D., 106n134
Sankey, H., 105n113
save the phenomena, 16, 19n25
Scarborough, J., 101n64
Schöner, H., 131n38
scientific explanation, 17, 19n26, 20n28, 53, 92, 103n101, 114; biomedical, 17, 26, 42, 47n55, 85, causation 29, 40, 80, generalist's knowledge 51, the individual and judgment 116, and pragmatism 96, redundant explanation, 53, standard regarding necessity 26, scientific

laws 27, 83, TE and ME 51–55, 59, 62, 63, 64, 65, 67, 71n25, n26, 72n38, 75n98, 76n123, 112; *see also* explanation-type α, explanation-type β, trial and error
Scythia, 7
Sebastian, A., 133n75, 134n82
Sedley, D., 21n33, 108n159
selected faith maxim, *see* epistemology, selected faith maxim
sequential causal accounts, *see* explanation-type α
Sextus Empiricus, 90, 105n119
Shakespeare, W., 20n28
Sharma, R., 100n28
Shaw, J., 23n93
Sherry, D., 105n114
similarity, 95
single perturbation account, 87, 88, 125
Sicilians, *see* Hippocratic writers, Sicilians
Singer, P.N., 133n58
skepticism, 6, 19n26, 50, 88–93, 132n45, radical, 3, 89, skepticism_1 def. 89, 91, 92, skepticism_2 def. 89, 91, 92, 98, skepticism_3 def. 89, 91, 92, 93, 95, 98, 105n113, 110, 135n96, *ataraxia* 89, 90, table interrelating categories 92; *see also* epistemology
Skeptics, 53, 88–93
skin, 59, 97, pores 12, 78, 113; *see also* respiration, skin breathing
Skinner, B. F., and Empiricists, 93
Smith, C.U.M., 72n38
Smith, W., 131n33
Socrates, 21n34, 45n29
Sophocles, 48n76
Sorabji, R., 71n25
Soranus, 25, 97, 108 n.153, n156–158, 129n1; and spontaneous abortion 97
species, 26, 51
sperm (*sperma*), 63, 65, 66, 67, 75n105
spinal cord, 41, 42
spiritedness; *see Thumos*
Steckerl, F., 80, 100n39
Sterelny, K., 44n15
Stoics, 14, 25, 112, 131n28, 132n46
stomach, 58
Strang, C., 74n82
Striker, G., 131n25
surgery, 16
Syennesis of Cyprus, 16
symptoms, 115, 116; *see* diagnosis
systematic philosophers, *see* philosophy, systematic
Szabó, A., 74n69

Tambiah, S.J., 19n16, 21n39
TE, *see* scientific explanation, TE
techne, see art
Tecusan, M., 108 n.154
teleology, 53, 55, 64, 65, 67, 71n25, 75n98, 76n123, 80, 109, 112; *see also* $\textit{phusis}_2$
testes, 66
Thales, 14, 23n89
Themison of Laodicea, 108n147
Theophrastus 10, 11, 22n73, 78
Theories of truth, pragmatism 96, correspondence 97
theory approach, *see* biomedicine, theory approach
Thessalos, 108n147
Thivel, A., 2, 19 n.15, n18
Thompson, G. 48n72
Thrämer, E., 102n78
Thumos, 15, 62, 118
Tieleman, T., 130n12, 131n34
treatment, 30, 38, 79, 98, 115, 116
Trepho, 39, 57, 65
trial and error, 39, 114
Trophe, 8, 37, 38, 56, 59, 63, 80, 125, 126
tuche, see chance
type-Token distinction, 26, 44n10, 49, 68, 76n119, 79, 92, 95, 98, 106n125, 109, 114, 116

ulceration, 97
upper preferred to lower, 72n54
uterus, 66, 75n113, 97, 98

vacuum, *see* nature, abhors a vacuum
Valtonen, M., 106n123
Van der Eijk, P., 18n10, 42n1, 45n23, 46n32, 99n7, n8, n10, n12, n18, 100n22, n26, n28, 100n33, 129n8, 132n47
Van Fraassen, B., 24n100
Van Woudenberg, René, 19n24
Vassale, C., 44n16
Vegetti, M., 106n129
veins, 80, 117, 118, thick 79, hepatic 112, emanated from left side of heart; *see also phlebes*
vena cava, 79, 130n18

venesection 23n98, 39, 126–128
Vermier, K., 20n27
Vesalius, V., 72n49, 88, 104n105
Vindicianus, 100n25
vivisection, 83–84
Vlastos, G., 23n91
Von Staden, H., 84, 93, 101n65, 102n81, 103n87, 106n130, n131

Wasserstein, A., 24n100
water, 41, 45n24, 59
Webb, O. 43n9
Wellmann, M., 25, 43n4, 46n34, 78, 99n12, n16, n18, 102n83
wet (as in hot, cold, wet, dry), *see* contraries
Whitehead, A.N., stand-alone systematic philosophy 110
Wilcox, M., 107, n137
Wilde, L., 48n76
Wilkie, J.S., 132n52, 134n80
Williams, George, 44n15
Wimsatt, William, 19n25
Winter, M., 44n18
women, *see* conception theory
Wood, R., 134n94
worldview, 2, 4, 6, 20n27, 132n53
Wright, T., 101n70, 103n98, 132n51

Xenophon of Cos, 79

Yahuda, A.S., 102n73

Zeno of Elea, 89
Zoroaster, 20n29

Texts Cited Index

Aeschines, *Epistulae*. 3.160, p. 18n6
Aeschylus: *Agamemnon,* II: 1478, p. 9, V. 1480, p. 33; *Eumenides* 658–660, p. 40
Anaxagoras, DK 14, p. 21
Anonymus Londinensis XXVI, 48c, XXVII, 6, 7, 38, p. 35; IV, 26–XXI, 9, p. 36; IV: 26–28, p. 37; XIV:16–32: 252–253, p. 43n6; 8, 10, 35, p. 100n33; 28.46–29.34, p. 134n90
Aristotle:
—*Categories* IV–IX, p. 51; 1a 20–1b 24; 2a 13–27, p. 71n32
—*Generation and Corruption* 336a, p. 52; 328b 26–338b 19, p. 132n48
—*Generation of Animals G.A.* 763b 31–33, p. 40; 743a 10–12, p. 57; 741b 15ff, p. 58; 765b, p. 60; 738a 13–14, 775a 18, 719a 30, 765b 15, p. 66; 719a, p. 66; 716a, p. 66; 729a 10, p. 67, 68; 730b 8–11, 19–23, p. 67; 729b 11, 729a 10, 728a 18, 737a 28, 732a 3–4, p. 68; 767b 21, p. 68; 741b 5, cf. 739b20, p. 71n24; 743a 10, p. 71n35; 726b 5, p. 72n47; 766 b 35, p. 73n57; 775a 17, p. 74n74; 744b 34, 86, 745b 18, 87, 737a 3, p. 74n85–7; 745b 15ff., 724b 26, p. 74n89–90; 726a 10, 100, 727a, 727b 30, 750b 11, 751a 4, p. 75n99–100; 725b 3, 728a 26, 728b 22, 737a 28, 750b 4, 766b 14, p. 75n101; 765b 5, 719b 2, p. 75n102; 747a 5, p. 75n105; 727a, 727b 30, 750b 11, 751a 4, 107, 746b 25ff., 108. 728b 34, 109. I.8; 746a 20, 734a 20 ff., 734a 28, p. 75n106–11; 739b 5, p. 75n113; 723b 28, 731a, 738b 25, 740b 25, 743a 25, p. 75n114; 768a 15 and 770a 6ff, p. 75n118; 768a 21–22, 766a 16–b16, 768a 14 vs. 768a 31, 768b 8, 768b 15, 766b 35, p. 76n120–2; 736b 8, 737a 17, p. 76n124; 732a 3–4, p. 76n126
—*History of Animals* 511 b 23–513 a 7, p. 16; 521a 2, a 18, p. 18n5; 511 b 23–513 a 7, p. 18n5; 561a 5, 9, 11, p. 58; 520b 12–22; 521a 6–9; 521a 18–21, p. 59; 514a 15, p. 61; 514a 16–22, p. 61; 521a 21–23, 26, 33, p. 65; 496a 22–27, p. 72n49; 3, 6, p. 72n51; 469a 4–19, p. 74n72; 495b 10, 496a 30, p. 74n94; 489a 24, p. 75n98; X, p. 75n101
—*Metaphysics,* 986a22, p. 9; 984a 5, p. 10; 1027a 20–21, p. 44n18; 980a 22, p. 53; 982a 5–983a 20, p. 70n19; 1071a 35–1074b 20, p. 70n19; I, 4, 8, p. 97; 983a 24–32, p. 131n21
—*Meteorology* 379b 18, p. 74n74; 340b 3, p. 74n75; 380a 11, p. 74n77; 380b 13, 381a 23, p. 74n78; 378b 20. 84, 379a, p. 74n83–4
—*Nicomachean Ethics* V, p. 45n21; 1094a 1, p. 53
—*On Dreams* 458b 9, p. 70n16
—*On the Heavens* 269b 31, 270b 23, p. 34; 288a 22–27, p. 70n19; III.2 301b 17, p. 70n19; 290a 32–33, p. 70n19

—*On Respiration* 473a 15, p. 12; 479b 19–22, 480a 3, 480a 17–19, p. 64; 481b 11, p. 71n23; 475a 25, p. 74n93; 480b, p. 74n94; 479b 32, p. 74n95; 480b 5, p. 75n86
—*On Sleeping and Waking* 455b 8, p. 70n16
—*On the Soul* 413a 7–9, p. 57; 429a 25–27, p. 57; 430a 22–24, p. 57; 413a 20–413b 2 ff, p. 71n30; 413a 4–9, 429a 24–27, 430a 21–25, p. 71n36; 413b 17, p. 76n125;
—*On Youth and Old Age* 469a 3–4, p. 74n92; 469b 8, p. 74n93
—*Parts of Animals* 651a 18, p. 18n4; 639a 1–7, p. 50; 639a 15–19, p. 51; 642a 32–642b 3, p. 54; 666a 2–22, 665a 33–35, p. 58; 656b, p. 61; 646a 25, p. 70n15; 660a 27–32, p. 70n18; 658a 9, p. 70n19; 642a 31–642b 4, p. 70n20; 642a 31–642b 4, p. 70n21; 640a 33–b 1, p. 71n25; 639b 25, p. 71n25; 668b 15, p. 71n35; 647b 10, 640b 20, 647b 22ff., 647b 13, 652a 25, 647b 18, p. 72n39–42; 647b 27, 651a 12, p. 72n45–6; 650b 10, p. 72n48; 651a, p. 72n50; 647b 35, p. 72n53; 647a 10, 658a 23, 665b 18, 665b 20, 686a 13, p. 72n54; 667a, 648a, p. 72 58n55; 648a 5, p. 73n58; 648a, p. 73n60; 650b 20, p. 73n61; 650b 24, p. 73n63; 652b 21–25, 653a 5, p. 73n65; 654a 32–b 13, p. 74n73; 651a 20, p. 74n88; 651a 17, p. 75n98; 689a 12, p. 75n101; 641a, 652b 15, p. 75n114; 646a 12–24, p. 132n48
—*Physics* 198b 35–36, p. 44n18; IV. 6–9, p. 47n50; 194b 24, p. 52; 194b 30, p. 52; VII, 1; VII, 4–5, p. 70n19; II 192b 20, p. 70n19; 194b 16–195b 30, p. 131n21
—*Politics* 1254b 13, 1260a 13, p. 67
—*Posterior Analytics*. 87b 19–22, p. 44n18; 73a 28, 73a 34, p. 50; II. 19, p. 50, 51; II. 1, p. 50; I.13, p. 50; 71b 9, p. 69n8; 72a 25, p. 70n9; 73a 21, p. 70n11; 83b 1, cf. 78a 28–78b 4, p. 71n27; 78a6–14, p. 71n28 ; 95a 9, p. 76n123 ; 94a 20–b 27, p. 131n21
—*Prior Analytics* II. 2–4, p. 71n28
—*Problems* 907a 18, p. 74n74
—*Rhetoric* 1370b 29–1372a 3, p. 71n29
—*Topics* I.15, 167b 1–3, p. 71n28
Arrian, vol. 1: *Anabasis Alexandri*, books i–iv; vol. 2: *Anabasis Alexandri*, books v–vii, p. 102n71
Augustine, *Against the Academicians*, Bk II, vols. 11–12; Bk III, vols. 23–26, p. 106n122

Celsus, *On Medicine*, 14–18, p. 82; 14–25, p. 83; Prooemium, 8, p. 100n31
Chalcidius, D–K, vol. 1, 212: 20–37, p. 46n35
Chaucer, *The Squires Tale*, 218, p. 20n28

Descartes: *The Passions of the Soul*: I.34–35, p. 22n55; I.10–11; 31–32, p. 118; *Discourse on Method*: 116–117, p. 44n19; *Meditations on First Philosophy,* II. 25–28, p. 104n109
Diocles, fr. 27, 51, van der Eijk, p. 99n19; fr. 74, 80, 101, 107.van der Eijk, p. 100n22; fr. 180–181, van der Eijk, p. 100n26
Diogenes, *Lives,* VIII, 30, p. 11

Empedocles: DK 98, p. 12; KR 392, 394, p. 13
Erasistratus: Fragments 195–196, Garofalo, p. 103n99; Fragments 33–43, Garofalo, p. 103n100
Euripides, *Orestes* 552, p. 40

Galen:
—*An in arteriis* K IV 712, p. 134n83; K IV, 713, p. 134n85; K IV 706–709, 733, p. 134n87
—*Car. VenSect*. 12—K XI, 288–289, p. 135n100
—*De anat. Adm*, p. 87; K. II 667–698, p. 130n17; K II 646ff, p. 134n87
—*De causis procatarcticis* VI, 59–70, p. 131n24

—*De differentiis pulsuum* 4, K VIII: 723, p. 101n58; IV.2, K. V: 561, p. 103n93
—*De difficultate respirationis* II, K VII 826, p. 113n, 30; II, K VII 826, p. 131n30
—*De methodo medendi* IX, 4, p. 126; IX, 10 K X, 363–68, 367–68, p. 135n100
—*De motu musculorum,* K IV 367–464, p. 130n17
—*De Placitis Hippocratis et Platonis,* Iwanis Mueller, 776, p. 115; K V, Bk I, p. 130n17; K V, Bk V (K V 568ff), p. 130n19; K V Bk VIII, 708, p. 131n36; K V 505–585, p. 133n61
—*De pulsibus ad tirones* K VIII, 474, p. 131n68
—*De sang in art,* p. 88
—*De sectis ingredientibus* K I, 64, p. 131n24
—*De symptomatum causis* K VII, 85–272, p. 134n81
—*De symptomatum differentiis* K VII, 47–48, p. 131n24
—*De trem. palp. convuls. et rig.* 6, K VII 614; 1, K VII: 584, p. 101n56; K VII, 584–642, p. 134n81
—*De usu partium* VI.10, p. 103n85; May, 89, 205, 207, 220, p. 130n19; K III, H I 340, p. 131n25; IV, 365–366, p. 131n35; VI.17 (May, p. 324), p. 133n62; K III, 4.9, K. III: 445, p. 133n64–5; VI. 21. 73, XV, 6, p. 133n72–3
—*De usu pulsuum* K V 166, p. 113; K V, 149, p. 120; KV, 163–165, p. 120
—*De usu resp.* I, K. IV: 471, p. 100n23; K IV 504–505, p. 123; KIV 5.4, 506, p. 131n29; K IV 504–505, p. 133n77
—*De Venet Art. Diss.* 9 Kühn II, 817, p. 72n49
—*Foet.Form.* VI 693, p. 131n35
—*Nat Fac* 3.4, p. 74n80; III. 15, p. 122; I. 16, p. 129n1; I. 77, p. 130n18; I. 2. 3 (K II 4–5, 7–9), p. 132n48
—*On Medical Experience* 13, 4, p. 99n3
—*Peri Spermatos* I.5, p. 70n19
—*Subfiguratio empirica,* chapts. 2, 6, p. 107n140; Ch. 8, p. 108n143

Harvey, W., *Exercitatio Anatomica de Motu Cordis et Anguinis in Animalibus*, 111, p. 104n106
Heraclitus, Fragments 197, 199, 203, 217, Kirk and Raven, p. 105n 115
Herodotus, History, 3.1, 3.129, p. 102n73; II.77, 3; II.84, p. 102n74
Herophilus, Fragment T 116, von Staden, p. 85; Fragment T248c, von Staden, p. 102n79; Fragment T79, von Staden, p. 103n89; Fragments 121, 80–85, von Staden, p. 103n90–2
Hippocrates (Hippocratic Writers)
—*Ancient Medicine* I, 1–4, pp. 31, 38; 24, pp. 45n30, 47n62; XVIII, 10–20, p. 74n79
—*Aphorisms* III. 28, p. 23n86
—*Breaths,* p. 38; I. 23–40, p. 47n63
—By Galen's account of Hipp. (K VI, 14), p. 113; *Diseases I.* 14, p. 39; *Diseases of Women* Littré VIII 130, 22, p. 46n39; *Airs, Waters, and Places,* p. 45n26; *The Oath,* p. 107n139; *On nature,* 12, line 1, p. 113
—*Epidemics I*, p. 30
—*Epidemics II*, Littré VII, 11, 26, p. 29
—*Epidemics III*, Littré VII, case 7, 160, p. 30; Littré VII, 123, p. 115
—*On Fleshes.* 11, p. 18; I, 1–3, 8; VIII: 584–88, p. 47n61
—*On Generation* IV, p. 23n86; Ch. 3–4, p. 41; 471, 474, ch. 4, p. 48n80–2
—*On Regimen* I. 10, p. 15; I.3, 1–11, p. 29; I, p. 45n24; I, 35, p. 60
—*On the nature of the Bones* (Littré IX, 174–176), p. 23n94
—*On the nature of the Child;* 493–494, p. 42; 492–496, p. 48n77
—*On the Wounds to the Head*, L. XXX, 19, p. 9
—*The nature of Man* I. 8–13, p. 14; (Littré) VI, 58–60, p. 23n88; (Littré) V, 1–6, 10, p. 29; 11, p. 42
—*The Sacred Disease* V, 1–3, p. 21; V, 1–5, p. 32; X. 20–25, p. 34

Homer (Homeric Writers)
—*Iliad*: I. 224–226, p. 7; V: 339–442, pp. 1, 75n98; X: 220, p. 8; XXI: 441, p. 8; XIX.111, p. 14
—*Odyssey:* IV: 572, p. 8; V: 389, p. 8; XVII.489, p. 22; IV: 220–232, p. 102n73
Hume, D., *Enquiry*, 16–20; 21–15; 49–53; 73–93, p. 20; 111–113, p. 104n108; 49–52, p. 107n141

Milton, J., *Paradise Lost,* p. 33
More, H *Enchiridion Metaphysicum;sive, de rebus incorporeia succincta & luculenta ; dissertation* I: 35–70, p. 44n19

Paley, W., *Natural Theology: Or Evidences of the Existence and Attributes of the Deity Collected from the Appearances of nature*: 1–13, p. 44n19
Papyrus Ebers, p. 85
Philolaus of Tarentum, p. 10; Fragment 1 Steckerl, p. 100n34
Phylotimus, Fragment 13 Steckerl, p. 101n50
Plato:
—*Laws,* 863 b, p. 130n9
—*Phaedrus,* 246a–257b, p. 74n68
—*Protagoras,* p. 45n29
—*Republic* 509d–511e, p. 45n28; IV, 439e–441b, p. 62; IV. 435–42, 9. 589d, p. 130n9; VI.504a, VIII.550b, IX.571b, p. 130n9; IV 435–442, p. 133n61
—*Theaetetu,s,* p. 129n4
—*Timeas* 83c, p. 18, 37; 91d 1–2, p. 40; 83c, p. 46n38; 83c 5 and 85c 3 and d 2, p. 72n52; 70b, p. 74n67; 44d, p. 74n70; 79 b&c, p. 100n24; 73b, p. 118; 69e, p. 130n9; 69–72, p. 133n61
Pliny:
—*Natural History* XXX, p. 20; XX, 19, 34, 52, 106, 139, 219, p. 100n26; XXV.5, 11–12, p. 102n73
—Plistonicus, Fr. 1, 4 Steckerl, p. 100n42
Plutarch, *Parallel Lives, Themistocles,* 29.5, p. 20
Praxagoras: Fragment 11, Steckerl, p. 80; Fragment 32, Steckerl, p. 81; Fragment 11, Steckerl, p. 82; Fragments 45,18, 21, 27, 32, 86, Steckerl, p. 100n29–30; Fragments 1, 98, Steckerl, p. 100n32; Fragments 70, 71, 27, 45, 24, 37, 45 Steckerl, p. 100n34–6; Fragments 7, 8, 9, 11 Steckerl, p. 101n48; Fragments 5, 9, 84, 85 Steckerl, p. 101n50; Fragment 3 Steckerl, p. 101n61

Sextus Empiricus, *Outlines of Pyrrhonism,* pp. 90–1
Shakespeare, *Winter's Tale* V.3.39, p. 20n28
Sophocles, *Antigone,* p. 48n76
Soranus, *The Gynecology,* 1.4. 19; 1.9.34; I.18.50, II.25 [45].52 [121]; II.25 [45].52 [121], p. 97; II. 25 [45]. 52 [121]; IV [XX]. 36 [85], pp. 97–8; III. 10.41, p. 108n157

Theophrastus, *History of Plants,* 1.12.2, 5.1.3, p. 74n77

Vesalius, *De humani corporis fabrica,* VII, cap. 19, p. 104n105

For Product Safety Concerns and Information please contact our EU representative GPSR@taylorandfrancis.com
Taylor & Francis Verlag GmbH, Kaufingerstraße 24, 80331 München, Germany

www.ingramcontent.com/pod-product-compliance
Ingram Content Group UK Ltd.
Pitfield, Milton Keynes, MK11 3LW, UK
UKHW022120190926
475702UK00013B/1556
* 9 7 8 0 3 6 7 8 6 8 4 3 7 *